W9-DIS-181

# *Choices*
## *Sex in the Age of STDs*

Second Edition

Jeffrey S. Nevid
*St. John's University*

Allyn and Bacon
Boston · London · Toronto · Sydney · Tokyo · Singapore

Photo Credits: Tony Neste, p.7; Bill Longcore/Science
Source/Photo Researchers, p. 26; Custom Medical Stock Photo,
p.66; E. Gray/Science Photo Library/Photo Researchers, p. 113.

Copyright © 1998, 1995 by Allyn & Bacon
A Viacom Company
160 Gould Street
Needham Heights, Massachusetts 02194

Internet: www.abacon.com
America Online: keyword: College Online

Library of Congress Cataloging-in-Publication Data


Nevid, Jeffrey S.
    Choices : sex in the age of STDs / Jeffrey S. Nevid. -- 2nd ed.
        p.   cm.
    Includes bibliographical references (p. ).
    ISBN 0-205-27829-9
    1. Sexually transmitted diseases--Popular works.  2. Safe sex in
AIDS prevention--Popular works.   I. Title.
    RC200.2.N483  1997
    616.95'1--dc21
                                                    97-22388
                                                       CIP

Printed in the United States of America

10 9 8 7 6 5 4 3 2    02 01 00 99 98

Portions of this material previously appeared in *A Student's Guide
to AIDS and Other Sexually Transmitted Diseases* by Jeffrey S.
Nevid, Copyright © 1993 by Allyn & Bacon.

# Contents

# FOREWORD

**CHOICES.** Never have the sexual choices we make carry such a heavy risk to our health and survival as they do today in the face of an epidemic of sexually transmitted diseases (STDs). Young people coming of age today face the threat of a deadly disease, AIDS, hanging over every sexual decision. AIDS—*acquired immunodeficiency syndrome*—may be the most prominent threat, but it is far from the only threat.

AIDS is a deadly viral condition in which the body is stripped of its ability to fend off life-threatening diseases. Despite recent treatment advances, AIDS very much remains a deadly disease. Lacking an effective vaccine, the best weapon we have against this killer disease is prevention. In this book you will learn what HIV infection and AIDS are, how HIV is spread, what it does to the body, how it is treated, and how it can be prevented. But AIDS is just one of many sexually transmitted diseases (STDs) that cause widespread suffering.

You are probably aware that AIDS is a killer. But how familiar are you with the risks associated with other STDs? Are you aware that chlamydia—one of the most prevalent STDs—can go undetected in the body for years, causing internal damage that can lead to infertility in women? Are you aware that the virus that causes genital warts is linked to cervical cancer in women and penile cancer in men? To put the risk of these other STDs in context, consider that HIV was found in only about 0.2 percent (1 in 500) of blood samples taken from a national sample of 16,861 students treated in university health facilities on 19 U.S. college campuses in the late 1980s. Yet health authorities believe that the microbes that cause chlamydia and genital warts infect one in ten college students.[1] This is not meant to downplay the threat of AIDS. AIDS is lethal, and one case is one too many.

New scientific information on AIDS and other STDs is reported at a dizzying pace. This book sifts through the welter of scientific findings to bring you the information you need to make responsible sexual decisions to protect yourself from sexually transmitted diseases. This book is meant to sound an alarm about the very real risks that STDs pose. While the choice of becoming sexually active rests with the individual, we hope that sexually active people will adopt responsible sexual practices that may lessen their risk of contracting or spreading these diseases. We hope that as knowledge is translated into responsible action, the epidemic of STDs we face today will become a relic tomorrow.

## Acknowledgments

The author wishes to acknowledge the contributions of the many scientists and clinicians whose efforts have begun to make inroads in the fight against this terrible epidemic of HIV infection and AIDS. The author wishes to acknowledge the assistance of the production team at Allyn & Bacon who helped to prepare this manuscript. The author is also thankful to Senior Editor Carolyn Merrill for her continuing support and belief in this book. The author also thanks Dr. Fern Gotfried who served as Consulting Medical Editor for the first edition of this book. The author also wishes to acknowledge that the views expressed in this book are not meant to represent those of any institutional affiliations.

## About the Author

**Jeffrey S. Nevid** is Professor of Psychology at St. John's University in New York, where he also directs the Doctoral Program in Clinical Psychology. He earned his Ph.D. in Clinical Psychology from SUNY Albany and holds a Diplomate in Clinical Psychology from the American Board of Professional Psychology. An active health researcher, Dr. Nevid was the recipient of a research grant from the National Heart, Lung, and Blood Institute of the National Institutes of Health which supported his work on smoking cessation with Hispanic smokers. He has published widely in a number of leading journals, including *Journal of Consulting and Clinical Psychology, Journal of Occupational Medicine, Journal of Youth and Adolescence, Journal of Family Counseling, Behavior Therapy, American Journal of Health Promotion,* and *Health Psychology*. He has also authored or coauthored numerous books including *Adjustment and Growth: The Challenges of Life, Abnormal Psychology in a Changing World, Human Sexuality in a World of Diversity, BT/Behavior Therapy, 201 Things You Should Know About AIDS and Other Sexually Transmitted Diseases,* and a forthcoming college text on personal health.

## CHAPTER ONE

# WHAT ARE STDs?

## *DID YOU KNOW THAT?*

- ▸ *Anyone who is sexually active can contract an STD, even you.*
- ▸ *Some sexually transmitted diseases (STDs) can also be transmitted nonsexually.*
- ▸ *A few STDs can be picked up by handling towels or bedding used by an infected person.*
- ▸ *If you are between the ages of 15 and 55, the chances are about one in four that you will contract an STD during your lifetime.*
- ▸ *You can greatly reduce your chances of contracting an STD by practicing the prevention strategies contained in this book.*
- ▸ *You can be infected with an STD and not realize it.*

HAROLD and CARIN, both 20, have been dating for several months. They feel strong sexual attraction toward each other but have hesitated to become sexually intimate because of fears about AIDS. Harold believes that using condoms is no guarantee against infection and wants the two of them to be tested for HIV, the virus that causes AIDS. Carin has resisted undergoing an HIV test, partly because she feels insulted that Harold fears that she may be infected, and frankly, partly in fear of the test results. She has heard that symptoms may not develop for years after infection. She wonders whether she might have been infected by one of the men whom she had slept with in the past.

STEFANIE has genital herpes. A 19-year-old pre-law student, she has had no recurrences since the initial outbreak two years earlier. But she knows that herpes is a lifelong infection and may recur periodically from time to time. She also knows that she may inadvertently pass the herpes virus along to her sexual partners, even to the man she eventually marries. She has begun thinking seriously about Steve, a man she has been dating for the past month. She would like to tell him that she has herpes before they become sexually intimate. Yet she fears that telling him might scare him away.

JOHN, 21, a math and computer science major, is planning a career in computer operations, hoping one day to run the computer systems for a large corporation. He lives off-campus with several of his buddies in a run-down house they've dubbed the "Nuclear Dumpsite." He has been dating Maria, a theater major, for several months. They have begun having sexual relations and have practiced "safer sex"—at least most of the time. During the past week he noticed a burning sensation while urinating. It seems to have passed now, so he figures that it was probably nothing to worry about. But he's not sure and wonders whether he should see a doctor.

Harold, Carin, Stefanie and John express some of the fears and concerns of a generation of young people who are becoming sexually active at a time when the threat of AIDS and other STDs (sexually transmitted diseases) hang over sexual decision. Coming of age in this age of AIDS, young people today face the prospect of a lifetime of uncertainty and fear overshadowing their development of intimate relationships. Here are a few comments from young people today about the threat posed by AIDS:[1]

ERICA, 18: I am terrified of AIDS. When my boyfriend cheated on me while drunk, I was scared. I now use condoms with him every time.

CAROL, 18: If you went to a bar and met someone before, you might say, "Let's go to your place," but now it's too big a risk because you don't know whom that person's been with. It's scary.

A 23-YEAR-OLD MAN: AIDS is a scary thing that always seems to be nagging away at the back of my mind. I've never been promiscuous, and I've never had casual sex. The frightening thing is that you can't be absolutely sure about your partner. With such a long incubation period, one mistake a long time ago can have fatal results.

RACHEL, 18: AIDS is a definite problem in college because you just don't know. At home, everyone knew who everyone slept with, but here you don't know who[m] they've slept with and they don't know who[m] you've slept with.

AIDS is indeed a scary thing, a very scary thing. But AIDS is only one of many STDs, albeit the most deadly and frightening. What are sexually transmitted diseases? How prevalent are they in our society? What are the risks of contracting an STD? And how can you protect yourself from contracting one?

**Sexually transmitted diseases** (STDs) are diseases that are transmitted through sexual means, such as by vaginal or anal intercourse, or oral sex. Until recently, the term *venereal disease (VD)*—after Venus, the Roman goddess of love—had been used to refer to diseases that are sexually transmitted. The term *venereal disease* was traditionally associated with two old menaces—gonorrhea and syphilis. Today we use the term *sexually transmitted diseases* (STDs) to refer to diseases that are spread by sexual contact. Still, you may find people (especially of your parents' and grandparents' generations) who continue to use the term venereal disease, or more simply, VD.

The term "sexually transmitted diseases" may cause some confusion, as some of these diseases may be spread (and often are) through nonsexual contact as well as sexual contact. For example, diseases such as AIDS and viral hepatitis may be spread among **injecting drug users** (IDUs) through the sharing of contaminated needles. And yes, a few STDs (like "crabs"—see Chapter 12) may be picked up from bedding or other objects, like moist towels, that harbor infectious organisms that cause these diseases or infestations.

> ## HEALTH TIPS
>
> ▸ Despite their name, some sexually transmitted diseases may be spread through nonsexual as well as sexual means.

# The Scope of the Epidemic

We are living through a period in which STDs are epidemic in our society and around the world. Many STDs are increasing at an alarming rate, especially among young people. Yet STDs are a threat to any of us, young or old, who are sexually active and don't know whether or not our partners are free of STDs. STDs can strike anyone—heterosexual or homosexual, young or old, black or white.

The U.S. has the unfortunate distinction of leading the world in the rate of STDs.[2] Overall, about one in four U.S. residents eventually contracts an STD[3] About one in twenty develop an STD each year.[4] About one in every six people is presently infected with an STD.[5] Nearly one-half of those infected with STDs are younger than 25 years of age.[6] Three million teenagers in the U.S. become infected with STDs annually.[7] The good news is that you can greatly reduce your risks of contracting (or spreading) STDs by practicing the prevention techniques described in this book.

> ## HEALTH TIPS
>
> ▸ Despite the prevalence of STDs in our society, you can greatly reduce your risk of contracting an STD by following the guidelines for prevention discussed later in the book.

The recent surge of STDs in our society reflects many factors. One is the increased numbers of young people who engage in coitus, many of whom practice "unprotected intercourse." More young people are having sex today and are becoming sexually initiated at younger ages than ever before.[8] According to most studies, the average person in the U.S. today engages in sexual intercourse for the first time between 16 and 17 years of age.[9] The increased rate of premarital sex among teens raises concerns about unwanted pregnancies and the risks of transmission of AIDS and other sexually transmitted diseases, especially since most teens fail to use condoms (which provide protection against both STDs and conception) regularly, if at all.

Another factor is that many people with STDs, especially those with chlamydia, have no symptoms and are unaware they are infected. Yet they can still pass along their STDs to others through engaging in unprotected sex.

Yet another factors is that couples often become sexually active without first discussing whether or not they should engage in sexual relations,[10] let alone discussing the necessary precautions they should take to protect themselves from unwanted pregnancies and STDs. Afterwards, they may describe their first sexual encounters as something that "just happened" or was an unplanned, "spur of the moment" experience. Yet this lack of communication may be largely responsible for the lack of preparedness that results in unwanted pregnancies and increased rates of STDs.[11]

The nearby "What Do You Say Now?" feature discusses how people can broach the subject of STDs with their partners before beginning sexual relations. Another factor indirectly contributing to the increased rate of STDs is the widespread use of birth control pills. Although birth control pills are reliable methods of contraception, they do not prevent the spread of STDs (although some people may assume that they do). Still another factor is that some STDs, such as chlamydial infections, are often symptom-free. Thus, people may not realize that they are infected and may innocently pass along the infection to others.[12]

Because some STDs are often symptom-free in their early stages, some readers have STDs today without realizing it. Ignorance is hardly bliss, however, because even symptom-free STDs may leave you sterile or damage your internal reproductive system or other organ systems if they are left untreated. STDs can also be painful and, in the cases of AIDS and syphilis, lethal. In addition to their biological effects, STDs exact an emotional toll on individuals and can strain relationships to the breaking point.

## HEALTH TIPS

▶ You can have an STD, even HIV infection, and not know it. Most people with HIV are unaware that they are infected. People with STDs can pass along the infection to others even if they are not aware of being infected.

# WHAT DO YOU SAY NOW?

## Talking to Your Partner about STDs

Many people find it difficult to discuss the issue of STDs with their partners.  As one young woman explained:[13]

> "It's one thing to talk about 'being responsible about STDs' and a much harder thing to do it at the very moment.  It's just plain hard to say to someone I am feeling very erotic with, 'Oh, yes, before we go any further, can we have a conversation about STDs?'  It's hard to imagine murmuring into someone's ear at a time of passion, 'Would you mind slipping on this condom or using this cream just in case one of us has an STD?'  Yet it seems awkward to bring it up any sooner if it's not clear between us that we want to make love."

Because talking about STDs with sex partners can be difficult or awkward, many young people admit that they "wing it."  They assume that their partners are free of STDs and hope for the best.  Some people pretend that if you don't talk about AIDS and other STDs, they will simply go away.  But the microbes causing AIDS, herpes, chlamydia, genital warts, and other STDs will not simply go away by not talking about them.  Then there is the case of a young man from rural Illinois who reported that when he was 16, his mother ordered him to buy condoms.  "Every weekend," he told an interviewer, "I had to show her that I had one with me or I would get grounded."[14]  The times, they are a-changing in this age of AIDS.

But just how might you raise the subject of STDs?  Timing is important.  As the young woman quoted above recognizes, it may be awkward to discuss STDs and the precautions you might take at a point in your relationship that it is not yet clear that you and your partner will become sexually intimate.  While you need not "jump the gun," whatever time you do select, make sure that it occurs *before* any genital contact begins (by that we mean genital-genital, oral-genital, or anal-genital contact).  Far too often, one partner will say to the other, *after* making love, something like, "By the way, I hope that you're not infected with anything, are you?"  Even more often, nothing at all is either said or done to prevent

STDs. Here are a few pointers that might make talking about STDs somewhat less awkward:[15]

**Photo 1.1 Talking to your partner about STDs.** The time to discuss STD prevention with your partner is before you engage in intimate sexual contact.

You've gone out with Chris a few times and you're keenly attracted. Chris is attractive, bright, witty, shares some of your attitudes, and, all in all, is a powerful turn-on. Now the evening is winding down. You've been cuddling, and you think you know where things are heading.

Something clicks in your mind! You realize that as wonderful as Chris is, you don't know every place Chris has "been." As healthy as Chris looks and acts, you don't know what's swimming around in Chris's bloodstream. Chris may not know either. In a moment of pent-up desire, Chris may also (How should we put this delicately?)... lie about not being infected or about past sexual experiences.

What do you say now? How do you protect yourself without turning Chris off? Write some possible responses in the spaces provided, and then check below for some ideas.

1. _____

2. _____

3. _____

Ah, the clumsiness! If you ask about condoms or STDs, it is sort of making a verbal commitment to have sexual relations, and perhaps you're not exactly sure that's what your partner intends. And even if it's clear that's where you're heading, will you seem too straightforward? Will you kill the romance? The spontaneity of the moment? Sure you might—life has its risks. But which is riskier: an awkward moment or being infected with a fatal illness? Let's put it another way: Are you *really* willing to die for sex? Given that few verbal responses are perfect, here are some things you can try:

1.  You might say something like, "I've brought something and I'd like to use it..." (referring to a condom).

2.  Or you can say something like "I know this is a bit clumsy," (you are assertively expressing a feeling and asking permission to pursue a clumsy topic; Chris is likely to respond something like "That's okay" or "Don't worry—what is it?") "but the world isn't as safe as it used to be, and I think we should talk about what we're going to do."

The point to this is that your partner hasn't been living in a remote cave. Your partner is also aware of the dangers of STDs, especially of AIDS, and ought to be working with you to make things safe and unpressured. If your partner is pressing for unsafe sex and is inconsiderate of your feelings and concerns, you need to reassess whether you really want to be with this person. You can do better.

**Some Common Signs and Symptoms of STDs**[16]

| In Men | In Women | In Men and Women |
|---|---|---|
| A drip or discharge from your penis | Unusual discharge or smell from your vagina | Sores, bumps, or blisters, near your sex organs, anus, or mouth |
| Pain in the testicles | Pain in your pelvic (lower belly) area or deep inside your vagina during sex | Burning or pain when you urinate |
| | Burning or itching around your vagina | Swelling or redness in your throat |
| | Bleeding from your vagina other than your regular menstrual period | Swelling in the area around your sexual organs |

*An important note.* Bear in mind that the symptoms listed in the table above may be due to other causes than STDs and that particular STDs may have different symptoms than those that are listed, or even no noticeable symptoms at all. The only way to know for sure is to obtain a medical evaluation. While our hope is to raise your awareness of the signs and dangers of STDs, this book is not intended to be used for self-diagnosis. If you experience any of the symptoms discussed in the text, or other unusual symptoms or changes in your physical condition, contact a physician immediately (that means today).

In the next few chapters you will learn about STDs caused by agents as diverse as bacteria, viruses, protozoa, and parasites. You will learn about their incidence, means of transmission, symptoms, diagnosis, and treatment. But more importantly, you will learn how your behavior can put you at risk of contracting these diseases, and what you can do to lower your risk. First, however, you can evaluate your beliefs and attitudes toward STDs to see whether they might heighten or lower your risk of contracting an STD.

## SELF-QUIZ
## STD ATTITUDE AND BELIEF SCALE

The following quiz can help you evaluate whether your beliefs and attitudes about STDs lead you to take risks that might heighten your risk of contracting an STD.

*Directions.* Choose whether the following items are true or false by circling the T or the F. Then consult the answer key that follows.

T F  1.   You can usually tell whether someone is infected with an STD, especially HIV infection.

T F  2.   Chances are that if you haven't caught an STD by now, you probably have a natural immunity and won't get infected in the future.

T F  3.   A person who is successfully treated for an STD needn't worry about getting it again.

T F  4.   So long as you keep yourself fit and healthy, you needn't worry about STDs.

T F  5.   The best way for sexually active people to protect themselves from STDs is to practice responsible sex.

T F  6.   The only way to catch an STD is to have sex with someone who has one.

T F  7.   Talking about STDs with a partner is so embarrassing that you're best off not raising the subject and hope the other person will.

T F  8.   STDs are mostly a problem for people who are promiscuous.

T F  9.   You don't need to worry about contracting an STD so long as you wash yourself thoroughly with soap and hot water immediately after sex.

T  F     10.  You don't need to worry about AIDS if no one you know ever came down with it.

T  F     11.  When it comes to STDs, it all in the cards.  Either you're lucky or you're not.

T  F     12.  The time to worry about STDs is when you come down with one.

T  F     13.  As long as you avoid risky sexual practices, like anal intercourse, you're pretty safe from STDs.

T  F     14.  The time to talk about safer sex is before any sexual contact occurs.

T  F     15.  A person needn't be concerned about an STD if the symptoms clear up on their own in a few weeks.

*Scoring Key*

1. *False*.  While some STDs have telltale signs, such as the appearance of sores or blisters on the genitals, or disagreeable genital odors, others do not.  Several STDs, such as chlamydia, gonorrhea (especially in women), internal genital warts, and even HIV infection in its early stages, cause few if any obvious signs or symptoms.  You often cannot tell whether your partner is infected with an STD.  Many of the nicest looking and well-groomed people carry STDs, often unknowingly.  The only way to know whether a person is infected with HIV is by means of an HIV-antibody test (see Chapter 5).

2. *False*.  If you practice unprotected sex and have not contracted an STD to this point, count your blessings.  The thing about good  luck is that it eventually runs out.

3. *False*.  Sorry.  Successful treatment does not render immunity against reinfection.  You still need to take precautions to avoid reinfection, even if you have had an STD in the past and were successfully treated.  If you answered true to this item, you're not alone.  About one in five college students polled in a recent survey of more than 5,500 college students across Canada believed that a person who gets an STD cannot

get it again.[17]

4. *False.* Even people in prime physical condition can be felled by the tiniest of microbes that cause STDs. Physical fitness is no protection against these microscopic invaders. think

5. *True.* If you are sexually active, practicing responsible sex is the best protection against contracting an STD (see Chapter 13).

6. *False.* STDs can also be transmitted through nonsexual means, such as by sharing contaminated needles or, in some cases, through contact with disease-causing organisms on towels and bed sheets or even toilet-seats.

7. *False.* Because of the social stigma attached to STDs, it's understandable that you might feel embarrassed raising the subject with your partner. But don't let embarrassment prevent you from taking steps to protect your own and your partner's welfare. Chapter 13 offers some guidelines that might help relieve embarrassment.

8. *False.* While it stands to reason that people who are sexually active with numerous partners stand a greater chance that one of their sexual partners carries an STD, all it takes is one infected partner to pass along an STD to you, even if he or she is the only partner you've had or even if the two of you only had sex once. STDs are a problem for anyone who is sexually active.

9. *False.* While washing the genitals immediately after the sex may have some limited protective value, it is no substitute for practicing responsible sex.

10. *False.* You never can know whether you might be the first one among your circle of friends and acquaintances to become infected. Moreover, symptoms of HIV infection may not appear for years after initial infection with the virus, so you may have sexual contacts with people who are infected but don't know it, and who are capable of passing along the virus to you. You in turn might then pass it along to others, whether or not you are aware of any symptoms.

11. *False.* Nonsense. While luck may play a part in determining whether you have a sexual contact with an infected partner, you can significantly reduce your risk of contracting an STD by following the guidelines in Chapter 13.

12. *False.* The time to start thinking about STDs (thinking helps, but worrying only makes you more anxious than you need be) is now, not after you have contracted an

infection. Some STDs, like herpes and AIDS, cannot be cured. The only real protection you have against them is prevention.

13. *False.* Any sexual contact between the genitals, or between the genitals and the anus, or between the mouth and genitals, is risky if one of the partners is infected with an STD.

14. *True.* Unfortunately, too many couples wait until they have commenced sexual relations to have "a talk." By then it may already be too late to prevent the transmission of an STD. The time to talk is before any intimate sexual contact occurs.

15. *False.* Several STDs, notably syphilis, HIV infection and herpes, may produce initial symptoms that clear-up in a few weeks. But while the early symptoms may subside, the infection is still at work within the body and requires medical attention. Also, as noted previously, the infected person is capable of passing along the infection to others, regardless of whether noticeable symptoms were ever present.

*Interpreting Your Score.* First add up the number of items you got right. The higher your score, the lower your risk. The lower your score, the greater your risk. A score of perhaps 13 correct or better may indicate that your attitudes toward STDs would probably decrease your risk of contracting them. Yet even one wrong response on this test may increase your risk of contracting an STD. You should also recognize that attitudes have little effect on behavior unless they are carried into action. Knowledge alone isn't sufficient to protect yourself from STDs. You need to ask yourself how you are going to put knowledge into action by changing your behavior to reduce your chances of contracting an STD.

# WHAT YOU SHOULD KNOW ABOUT THE AIDS EPIDEMIC

## *DID YOU KNOW THAT?*

- *HIV, the virus that causes AIDS, can strike anyone, even young vigorous heterosexuals.*
- *AIDS kills by disabling the body's defenses against life-threatening illnesses.*
- *In 1990, one death from AIDS occurred every 12 minutes in the United States.*
- *Heterosexual transmission of AIDS accounts for three of four AIDS cases worldwide.*
- *AIDS is increasing more rapidly among women than men in the U.S.*
- *Your behavior, not the groups to which you belong, determines your risk of contracting HIV.*

In 1981, Michael Gottlieb was the first physician to describe the effects of AIDS in the medical journals. His experience with AIDS began when he was treating a number of male patients who were suffering from *Pneumocystis carinii pneumonia* (PCP). *Pneumocystis carinii pneumonia* is a rare form of pneumonia. Until that year, PCP had been typically found among cancer patients whose immune systems were suppressed as a side effect of chemotherapy. Some of the PCP sufferers showed other disorders that were associated with suppressed immune systems—high fevers, weight loss, and *candidiasis* of the mouth, which is a kind of yeast infection.

Normally speaking, patients with these disorders recovered, and Gottlieb expected that these men would as well.[1] It was not to be, however. All of them died.

In the early 1980s, physicians also began to see cases in young men of a rare form of cancer—*Kaposi's sarcoma*—that leaves purple spots on the body. This illness usually struck only aging Jewish and Italian men. While aging men usually live with the disease and die later of other causes, the young men quickly deteriorated and died. Some also had PCP.

These men and those examined by Gottlieb had one thing in common: they were all gay. The syndrome without a name was soon dubbed the "gay plague" or the "gay cancer" and officially became known as GRID, for *gay-related immune deficiency*. Social critics contend that the government initially responded slowly to the epidemic because of prejudice against homosexuals.[2] Soon, the so-called gay plague began to strike heterosexuals, mostly IDUs and the sex partners of IDUs and bisexual men—as well as babies born to infected mothers and hemophiliacs and others who received contaminated blood transfusions.[3] GRID then became known as AIDS, as the disease could no longer be classified as a syndrome that beset gay men only. AIDS threatens us all, whether we are straight or gay, rich or poor, black or white, male or female, or young or old.

# What is AIDS?

AIDS is a syndrome, or cluster of symptoms. It is caused by a virus, the *human immunodeficiency virus* (HIV), that invades and then disables the body's natural line of defense against disease, the immune system, leaving the person vulnerable to various infections and diseases. As AIDS progresses, the body becomes less capable of fending off serious illnesses that are normally held in check by a healthy immune system. Because these diseases take advantage of the body's weakened immune system, they are dubbed opportunistic diseases. AIDS itself does not kill, but eventually the person succumbs to one of these opportunistic diseases or infections.

HIV infection typically does not progress to full-blown AIDS for about ten years following initial infection, on the average. During that time, the person may appear healthy or have symptoms of HIV infection that do not yet merit a diagnosis of full-blown AIDS.

As far as we know, once you get HIV you've got it for life. Recent medical advances have raised hopes that a combination of drugs *may* be able to cure the disease by eradicating the virus from the body.[4] Yet these are hopes, not certainties. However, doctors are increasingly optimistic that antiviral drugs used in combination may turn AIDS into a manageable chronic disease, similar to diabetes, rather than the terminal illness or automatic death sentence we've known it to be (see Chapter 5).[5] Doctors have also made substantial progress in treating some of the infections that take hold in AIDS patients whose weakened immune systems leave them vulnerable to a host of opportunistic diseases.

---

### HEALTH TIPS

▶ Recent progress in the war against AIDS should not lull you into a false sense of security. AIDS remains a deadly disease. Don't bank on the possibility of a future cure. Your best protection against HIV is to prevent it from getting into your body.

---

### Where Did the AIDS Virus Originate?

Although no one knows where HIV originated, it is suspected that it may represent a variant of viruses found in monkeys and chimpanzees in Africa that somehow crossed over to humans.[6] Scientists have clearly established that a similar virus exists in a West African monkey.[7] Scientists have discovered several types of HIV that cause AIDS, including *human immunodeficiency virus type 1* (HIV-1), the most prevalent form, and a more recently discovered virus, *human immunodeficiency virus type 2* (HIV-2) . HIV-1 appears to be the more virulent form of the virus.[8]

Some scientists suspect that the virus was introduced in humans as the unintended result of peculiar malaria experiments dating from the 1920s into the 1950s. In those experiments, people were inoculated with blood from monkeys and chimpanzees that may have been infected with the viral ancestors of HIV.[9] The experiments were intended to determine whether malaria parasites in the primates would infect people. Further evidence is needed to substantiate the role, if any, that these experiments may have played in the transmission of HIV-type viruses to humans.

## How Prevalent is HIV/AIDS?

Fewer than 100 Americans had died of AIDS in 1981, the year that Gottlieb first reported the occurrence of a cluster of symptoms that would later be identified with AIDS.[10] By 1997, nearly 600,000 Americans would be diagnosed with AIDS. More than 350,000 would die from it. In 1995 alone there were more than 74,000 new cases of AIDS in the United States.

By the early 1990's, one American was dying from AIDS every 12 minutes.[11] The prevalence of AIDS increased steadily in the early 1990s at a rate of about 3 to 5 percent a year, to about 62,000 new cases per year. AIDS is increasing most rapidly in our society among women, people of color, people who use injectable drugs, and people who engage in unprotected, heterosexual sex. In 1996, Blacks, for the first time, accounted for more new cases of AIDS than Whites, 41% versus 38%.[12] Hispanics account for another 19% and other races for 2%.

By 1996, an estimated 700,000 people in the U.S. were believed to be infected with HIV.[13] The World Health Organization (WHO) estimated that 20 million people around the world are infected. Nearly 4.5 million of them had developed AIDS.[14] WHO estimates that the number of persons infected with HIV may soar to 30 to 40 million by the year 2000.[15] An even grimmer outlook is projected by AIDS experts and researchers at the prestigious International AIDS Center at Harvard University. The Harvard researchers estimated that by the year 2000, as many as 110 million people worldwide might be infected.[16]

The Harvard group charged that too little is being spent to fight the epidemic, especially in developing countries. But nowhere in the world has the epidemic yet peaked. The rate of HIV infection is increasing faster in many developing countries in Asia, Latin America, and Africa than in the United States and other industrialized nations. Sub-Saharan Africa has been most severely hit by the AIDS epidemic. This region accounts for nearly 70% of the cases of HIV infection and AIDS cases around the world.[17] Asia now ranks as the continent with the second greatest number of cases of HIV infection, after Africa. The crisis is most severe in India and Thailand. As in Africa, male-female sexual contact is the major route of transmission in Asia.

# Who is Most at Risk?

In the United States, AIDS remains a disease that predominantly affects men who have sex with other men or who inject drugs.[18] By 1996, about 40% of new cases involved male-to-male sexual transmission, while about 25% involved needle-sharing among injecting drug users.[19] The rates of transmission among gay men have been declining while they have rising among people who inject drugs *(injecting drug users* or IDUs) and their sexual partners. Figure 2.1 shows the geographic distribution of AIDS cases within the United States.

The declining rate of HIV infection among gay males is largely attributed to increased use of safer sexual practices, including increased use of condoms and reductions in the numbers of sexual partners.[20] Experts remain concerned, however, about resumptions (relapses) of unsafe sexual practices in gay men and about younger gay men who may not follow safer sex guidelines.[21]

IDUs transmit the virus by sexual contact and by sharing contaminated hypodermic needles. In New York City, nearly nine of ten cases of heterosexual transmission of HIV involve IDUs and their (non-drug using) sex partners.[22]

Some communities, especially inner-city neighborhoods with widespread injecting drug use, have been especially hard hit. AIDS cuts across all racial, ethnic, occupational, and social class groupings, however.[23]

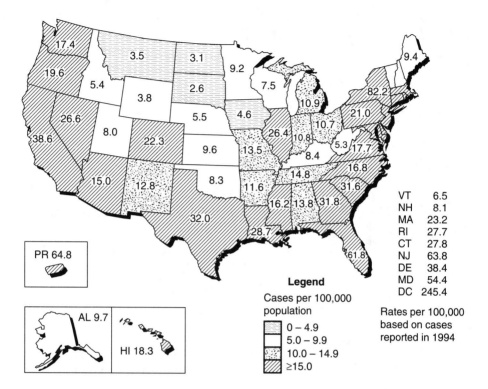

**Figure 2.1 AIDS Cases Within the United States**. Source: Centers for Disease Control and Prevention (1995). *HIV/AIDS Surveillance Report*. Year-end Edition, Vol. 6, No. 2.

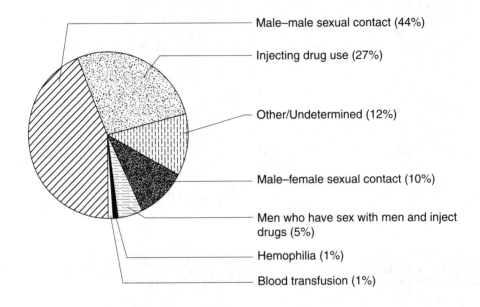

Male–male sexual contact (44%)

Injecting drug use (27%)

Other/Undetermined (12%)

Male–female sexual contact (10%)

Men who have sex with men and inject drugs (5%)

Hemophilia (1%)

Blood transfusion (1%)

Data reflect cases reported in 1994

**Figure 2.2  AIDS Cases by Exposure Category**. Source: Centers for Disease Control and Prevention (1995). *HIV/AIDS Surveillance Report*. Year-End Edition, Vol. 6, No. 2.

## Heterosexual Transmission in U.S. is Increasing

Heterosexual transmission via sexual intercourse accounts for most cases of HIV transmission in sub-Saharan Africa, the Caribbean, and in some areas of South America.[24] Worldwide, heterosexual transmission via sexual intercourse accounts for 75 percent of cases of HIV infection.[25] In the U.S., heterosexual contact is the fasting-growing exposure category and now accounts for about 10% of cases of AIDS (see Figure 2.2). Among women, 40% of the new cases of AIDS reported between July 1995 and June 1996 were attributed to male-female sexual contact. The number of new cases of AIDS in men and women attributable to heterosexual contact more than doubled between 1989 and 1992.[26]

Although heterosexual spread of HIV in the U.S. has remained largely focused within the community of IDUs and their sex partners,[27] health officials warn that unless prevention efforts take hold, HIV is likely to spread more widely within the general heterosexual population in the U.S., as it has in other parts of the world.

Despite increased talk lately about "safer sex" techniques, such as the use of latex (rubber) condoms to prevent the sexual transmission of HIV and other infectious organisms, the message does not seem to be getting across to the majority of young people in the U.S. and Canada.  A survey of more than 5,500 college students on 51 Canadian campuses revealed that while most students were generally knowledgeable about HIV transmission, most of the sexually active students said they never or only occasionally used condoms. Only 25 percent of the men and 16 percent of the women always used a condom during sexual intercourse.  Yet nearly half of the men and women polled reported engaging in numerous sexual encounters with different partners.[28]  Fewer than half of a sample of New England college students had changed their sexual behavior because of the advent of AIDS.[29]  Despite growing awareness of AIDS, a survey of college women who attended a student health service in the northeast in the 1970s and 1980s found little dropoff in the incidence of coitus with multiple partners.[30]  Nor did most of the sexually active women who attended this clinic reliably use condoms.  Another survey of University of Massachusetts students revealed that some 70 percent had *not* changed their sexual practices in any way, despite the threat of AIDS.[31]  Another survey found that only 35 percent of sexually active students at two southeastern state universities used condoms reliably.[32]  A survey in an upstate New York college showed that only one in five reported always using condoms.[33]  Yet another survey of southern college students found that half did not think of themselves as being at risk of contracting HIV.[34]

## AIDS is Increasing More Rapidly Among Women

Although AIDS was once considered a syndrome that afflicted only gay men, AIDS cases are rising faster among women than among men. In the mid-1990s, the number of women with AIDS in the United States was doubling every year.[35] The numbers of reported AIDS cases among women may represent the tip of iceberg, because symptoms may not develop for years after infection with HIV. Approximately one in four women with AIDS is 20 to 29 years of age. Given the latency of progression to full-blown AIDS, many of these women were likely infected as teenagers.[36]

## Your Behavior Determines Your Risk of HIV Infection—Not the Groups to Which You Belong

It is your behavior and not the groups to which you belong that determines your relative risk of infection.  As a result, many investigators of the epidemiology of AIDS now focus on high-risk *behaviors* (such as unprotected intercourse) rather than high-risk *groups*.[37]  It is not true that only people in traditionally identified high-risk groups are at significant risk of contracting HIV.

| HEALTH TIPS |
| --- |
| ▸   It is one's behavior, and not the groups to which one belongs, that places one at risk of contracting HIV. |

# What Does the Future Hold?

Despite progress in treating HIV infection and AIDS, the epidemic is far from over. What the future holds remains uncertain. The question you need to ask yourself as you go through this book is, "What can I do to protect myself from becoming yet another statistic of the epidemic of AIDS and other STDs?"

## CHAPTER THREE

# WHAT YOU SHOULD KNOW ABOUT HIV/AIDS ILLNESS AND TRANSMISSION

## *DID YOU KNOW THAT?*

- *HIV attacks and disables the very cells that the body normally relies on to fight off foreign agents, including HIV itself.*
- *AIDS does not kill directly; rather it kills by disabling the body's ability to fend off other life-threatening illnesses.*
- *Most people with HIV infection feel perfectly fine and may have no symptoms for years following initial infection.*
- *People can pass along HIV to others even if they have no symptoms of the infection themselves.*
- *The average length of time between HIV infection and the development of full-blown AIDS is about ten years.*
- *The life expectancy of the average person diagnosed with AIDS is a little more than a year.*

How much do you know about transmission of HIV, the virus that causes AIDS? Can HIV be transmitted by vaginal intercourse? By a handshake with an infected person? By sharing utensils or a drinking cup? By deep "French kissing"? By a mosquito bite? In order to protect yourself, you should understand how HIV *is* and *is not* transmitted.

AIDS is a killer, but how does it kill? What cells in the body does HIV attack and destroy? What role do these cells play in the body's natural defenses against disease?

How much do you know about the progression from HIV to AIDS? How long is the average person infected with HIV before he or she develops AIDS? How likely is it that a person will progress from HIV to AIDS?

# AIDS and the Immune System

AIDS is caused by a virus that attacks the body's immune system—the body's natural line of defense against disease-causing organisms. To understand the effects of the AIDS virus, we must first know something about the components and functioning of the immune system. Given the intricacies of the human body and the rapid advance of scientific knowledge, we tend to consider ourselves dependent on highly trained medical specialists to contend with illness. Actually we cope with most diseases by ourselves, through the functioning of our **immune systems.**

The immune system combats disease in a number of ways. It produces a trillion white blood cells that systematically envelop and kill **pathogens** like bacteria, viruses, and fungi; worn out body cells; and cells that have become cancerous. White blood cells are referred to as **leukocytes.** Leukocytes engage in microscopic warfare. They undertake search-and-destroy missions; they identify and eradicate foreign agents and debilitated cells.

Leukocytes recognize foreign agents by their surface fragments. These surface fragments are called **antigens** because the body reacts to their presence by developing specialized proteins, or **antibodies,** that attach to the foreign bodies, inactivating them and marking them for destruction. (Infection by HIV may be determined by examining the blood for the presence of specific antibodies to the virus. Unfortunately these antibodies are unable to eradicate the infection.)

Special "memory lymphocytes" (lymphocytes are a type of leukocyte) are held in reserve, rather than marking pathogens for destruction or going to war against them. They can remain in the bloodstream, sometimes for several years, and they form the basis for a quick immune response to an invader the second time around.[1]

## HIV Attacks CD4 Cells,the "Quarterback" of the Immune System

HIV directly attacks the immune system by invading and destroying a particular type of lymphocyte called the CD4 lymphocyte (also called the $T_4$ cell or helper T-cell), the quarterback[2] of the immune system (see Figure 3.1). Helper T-cells "recognize" invading pathogens and signal B-lymphocytes or B-cells—another kind of white blood cell—to produce antibodies that inactivate pathogens and designate them for annihilation. Helper T-cells also signal another class of T-cells, called killer T-cells, to destroy infected cells. By attacking and destroying helper T-cells, HIV disables the very cells that the body relies on to fight off this and other diseases.[3] As HIV cripples the body's natural defenses, the individual is exposed to serious infections and cancers that would not otherwise take hold. When the immune system is disabled, these diseases proliferate and prove difficult to control, and eventually result in death.

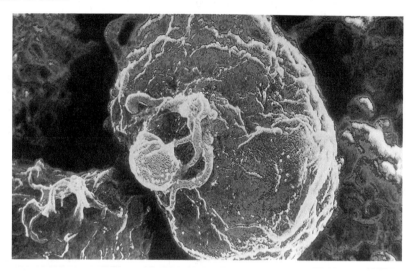

**Figure 3.1  HIV Attacking a White Blood Cell**

### HIV Gradually Reduces the Number of CD4 Cells

The blood normally contains about 1,000 CD4 cells per cubic millimeter.[4] The numbers of CD4 cells may remain at about 1,000 per cubic millimeter for several years following infection, and many people show no symptoms while they remain at this level. Then the numbers of CD4 cells decline, although many people still show no symptoms for several more years. People become most vulnerable to diseases when the level of CD4 cells falls below 200 per cubic millimeter.[5]

# The Progression of HIV Infection to AIDS

HIV follows a complex course once it enters the body. Many adults are symptom-free for years. In fact, most people today with HIV infection feel pretty well and are not under a physician's care.[6] Most are not even aware that they harbor the deadly virus. Shortly following infection, the person may experience mild flu-like symptoms, however—fatigue, fever, headaches and muscle pain, lack of appetite, nausea, swollen glands, and possibly a rash. Such symptoms usually disappear within a few weeks, and the person may remain symptom-free for months or years. People may thus dismiss these symptoms as a passing case of flu. People who enter this asymptomatic or carrier state may look and act well, not even realizing that they are infectious. Thus, they can unwittingly pass along the virus to others.

---

## HEALTH TIPS

▸ You cannot tell whether you are HIV infected on the basis of symptoms. The early symptoms of HIV infection may involve flu-like symptoms that can also be due to a host of other causes. In some cases, symptoms of HIV infection may not occur for months or even years after initial exposure.

---

## HIV Symptomatic State

Some HIV-infected people remain asymptomatic carriers for periods of months or years. Others enter a symptomatic state that is typically denoted by such symptoms as chronically swollen lymph nodes and intermittent weight loss, fatigue, fever, and periods of diarrhea. The severity of symptomatic HIV infection depends on various factors, such as the person's general health. This symptomatic state does not constitute full-blown AIDS, but it indicates that the disease process is taking a toll on the person's immune system. People with symptomatic HIV infection may mistakenly attribute their symptoms to other causes and not realize they are infected with HIV. Like asymptomatic carriers, they may unknowingly pass along the virus to others through sexual contact or needle-sharing.

## Progression to Full-Blown AIDS

Eventually, perhaps a decade or more after the person is initially infected with HIV, and for reasons that remain unclear, the virus begins to propagate in large numbers. It obliterates the cells that house it and spreads to infect other immune-system cells, eventually destroying or disabling the body's ability to defend itself from disease. About half of the people with HIV develop diagnosable AIDS within 10 years of initial infection.[7] Yet the length of time between HIV infection and the development of AIDS varies greatly among individuals. Some people develop AIDS within a year of infection with HIV, while others do not progress to AIDS for more than 10 years. For this reason, people infected with HIV may feel as though they carry time bombs within them.

AIDS is classified as a *syndrome*, or cluster of symptoms, because it is characterized by a variety of different symptoms. The beginnings of full-blown cases of AIDS are often marked by such symptoms as fatigue, "night sweats," persistent fever, swollen lymph nodes, diarrhea, and weight loss that cannot be accounted for by dieting or exercise.

## How is AIDS Diagnosed?

The diagnosis of AIDS is based on the appearance of various indicator diseases, such as PCP; Kaposi's sarcoma; toxoplasmosis of the brain, which is an infection of parasites; herpes simplex with chronic ulcers; or wasting syndrome (wasting away, that is, as in losing weight without notable dietary changes or expenditure of calories through exercise). These diseases are all termed **opportunistic diseases**. People are not likely to fall victim to them unless their immune systems are weakened and unable to fend them off. Hence the presence of these signs in a person who is HIV-infected is taken as evidence of full-blown AIDS.

In 1992, the federal Centers for Disease Control expanded the diagnostic criteria for AIDS. Three additional diseases were added to the list of 23 other AIDS indicators: tuberculosis of the lungs, recurrent pneumonia, and invasive cervical cancer. These added conditions commonly affect HIV-infected IDUs and women. In addition, the new criteria provides for an AIDS diagnosis when the CD4 cell count in persons infected with HIV falls to fewer than 200 cells per cubic millimeter, about one fifth the normal amount, irrespective of the presence of indicator diseases.

The more inclusive definition makes more people eligible for AIDS benefits. It is also intended to better accommodate women infected with HIV, who may develop gynecological conditions not seen in men, such as persistent and recurrent vaginal yeast infections, as well as IDUs, infected children, and unusual cases in which the traditional indicator diseases are not present.

## Outlook for Persons with AIDS

Only a few short years ago the outlook for AIDS patients was bleak. As AIDS progressed, the person grew thinner and more fatigued, became unable to perform ordinary life functions, and fell prey to opportunistic infections, leading within a few months or a few years to death. Then came reports in 1996 of AIDS patients making miraculous recoveries, some literally from death's door. The year 1996 marked a watershed year in the fight against AIDS. That year marked the introduction of a new class of drugs, the protease inhibitors, which used together with other antiviral drugs, was capable of reducing the levels of HIV in the blood to below detectable levels in many cases.[8] Whether these advances in treatment foretell a future victory against this terrible disease remains uncertain. But at least now we have something on which we can pin our hopes and dreams.

# What You Should Know about HIV Transmission

HIV can be transmitted by contaminated bodily fluids (blood, semen, or vaginal secretions), which enter the body as the result of vaginal, anal, or oral-genital intercourse with an infected partner, sharing a hypodermic needle with an infected person (as is common among IDUs), transfusion with contaminated blood, transplants of organs and tissues that have been infected with HIV, artificial insemination with infected semen, or being stuck by a needle used previously on an infected person.[9] Let us take a closer look at the ways in which HIV *is* and *is not* transmitted.

## Sexual Transmission

HIV may be transmitted from one partner to another through vaginal or anal intercourse, oral-genital sex, or other intimate sexual contacts that involve contact with the bodily fluids of an infected person. HIV may enter the body through tiny cuts or sores in the mucosal lining of the vagina, rectum, and even the mouth.[10] These cuts or sores may be so tiny that you may not be aware of them.

---

### HEALTH TIPS

▸ HIV can be transmitted through various forms of intimate sexual contact, including vaginal or anal intercourse and oral-genital sex.

---

Sexual activities may become a means of transmission of HIV only if one of the partners is infected with the virus. *You cannot contract or transmit HIV via sexual activity if neither you nor your partner is infected, no matter what sexual activities you practice.* This is not to say that these activities are entirely free of risk.

They may serve as a mode of entry for other STD-causing microorganisms, such as those that cause syphilis, gonorrhea, genital warts, or chlamydia. In addition, anal intercourse may cause injury to sensitive rectal tissue if it is performed too forcefully or without sufficient lubrication.

---

### HEALTH TIPS

▸ You cannot contract HIV from an uninfected partner no matter what sexual activities you practice. Yet sexual practices with non-HIV-infected partners may incur other risks, such as the spread of other STDs, or physical injuries to sensitive rectal tissue caused by forceful anal penetration.

---

While it is important to separate fact from fiction regarding the means of transmission of HIV, you should not get the impression that unprotected sexual activities are safe from HIV transmission unless you know that your partner is infected with HIV. Rather, the reverse is true: *You should consider unprotected sex to be unsafe unless you know (not guess, know!) that your partner is uninfected.*

---

### HEALTH TIPS

▸ Unless you and your partner have been continually monogamous for a period of years and are known to be free of STDs, you should consider any form of unprotected sex to be risky.

---

**Male-to-Male Transmission**   In the U.S., sexual contact between gay and bisexual men, primarily through anal intercourse, has been the primary mode of transmission since the virus first appeared. Transmission may occur through either the receptive (insertee) or inserter role, although the receptive role poses the greater risk. The lining of the rectum provides a convenient port of entry into the bloodstream for HIV, as it is very thin and tears easily, allowing the virus to penetrate into the bloodstream through these tiny tears. (The vaginal lining is much thicker and more resistant to tearing). These tiny tears may occur even in the absence of any noticeable blood. Engaging in unprotected anal intercourse is perhaps the most risky sexual practice.

The risk of transmission of HIV through anal intercourse is not limited to gay or bisexual males. Heterosexual couples are at just as great a risk from practicing unprotected anal intercourse if one of the partners is infected as are gay male couples. Remember, it's one's behavior and not the groups to which one belongs, that determine one's risk of HIV infection.

**Heterosexual Transmission** Heterosexual transmission of HIV via vaginal intercourse occurs more often from male to female than from female to male,[11] partly because more of the virus is found in the ejaculate than in vaginal fluids. About two of three cases of U.S. heterosexual transmission involve male-to-female transmission.[12] A man's ejaculate may also remain for several days in the vagina, providing greater opportunity for infection to occur. HIV transmission from woman-to-woman is possible, but much less likely than male-to-female transmission.[13]

---

## HEALTH TIPS

▶  Although heterosexual transmission of HIV is more likely from male-to-female than the reverse, both men and women can become infected with HIV through heterosexual sex.

---

**Transmission of HIV through Oral Sex** Evidence has accumulated that shows that HIV may also be transmitted by oral sex (oral-genital contact involving mouth-on-penis or mouth-on-vagina) with an infected partner, regardless of the gender of the partner.[14]

---

### HEALTH TIPS

▸ Although oral-genital sex appears to be a less likely mode of transmission of HIV than either vaginal or anal intercourse, it is best to avoid unprotected oral sex with a partner who is not known to be free of HIV.

---

Though scientists suspect that transmission of HIV through deep kissing, prolonged kissing, or "French" kissing is theoretically possible, this mode of transmission is considered unlikely.[15] We have yet to find any confirmed cases of HIV transmission through kissing. Still, it may be wise to avoid deep, open-mouth kissing with partners who are not known to be uninfected.

---

### HEALTH TIPS

▸ Although the role of "deep" kissing in transmitting HIV remains unclear, it would be safer to avoid deep, open-mouth tongue kissing with partners who are not known to be uninfected.

---

**Factors Increasing the Risk of Sexual Transmission** Some people are apparently more vulnerable to infection by HIV than others. As a general rule, the probability of transmission rises with the number of coital contacts with an infected partner. Yet there is no predictable connection between the number of episodes of unprotected sex with an infected person and the probability of transmission.[16] Some people seem more likely to transmit the virus, and others seem to be especially vulnerable to contracting it. Why, for instance, are some people infected by one sexual contact with an infected partner, whereas others are not infected by months or years of unprotected coitus?

Some clues have begun to emerge, based on studies in the United States, Europe, and Africa. A history of STDs may heighten the risk of infection by HIV. STDs like genital warts, gonorrhea, trichomoniasis, and chlamydia inflame the genital region, which may heighten the risk of sexual transmission. STDs that produce genital ulcers, like syphilis and genital herpes, may heighten vulnerability to infection by allowing HIV to enter the circulatory system through the ulcers.[17] So if you are infected with one of these STDs, you may stand an increased chance of contracting HIV by engaging in sexual contact with an infected person, especially if you engage in unprotected sex.

The probability of transmission also appears to be affected by such factors as one's sexual practices, the amount of HIV in the semen, and circumcision.[18] Anal intercourse, for example, may provide a convenient port of entry for HIV by tearing or abrading the tissue of the rectal lining. It also appears that the amount of virus in semen varies throughout the course of HIV infection and AIDS, reaching peaks shortly after infection and when full-blown AIDS develops. Circumcised men may have a lower risk of infection because they are less likely to have genital ulcers. Moreover, HIV cannot accumulate under the folds of the foreskin in men who have been circumcized.[19] Cells in the foreskin may be particularly vulnerable to HIV infection as well.[20] More research is necessary to confirm the linkage between circumcision and HIV infection, however.

## Transmission of HIV through Injecting Drug Use

As noted in Chapter 2, people who use injecting drugs (IDUs) account for about one in four AIDS cases in the U.S. The rate of HIV infection among IDUs and their sexual partners has also been increasing. IDUs risk infection through sharing contaminated needles when using drugs and through practicing unsafe sex. When a person shoots drugs, a small amount of their blood remains inside the needle and syringe. If the person is HIV-infected, the virus may be found in the blood remaining in the needle or syringe. When others use the same needle, they directly inject the infected blood into their bloodstreams.[21]

AIDS-prevention programs have apparently decreased needle sharing and, occasionally, drug injection altogether, but they seem to have little impact on changing risky sexual behavior in this population.[22] In one study of a group of methadone patients, slightly more than half (54%) reported they had adopted safer injection practices, but only 14 percent reported engaging in safer sexual practices.[23]

HIV may also be spread by sharing needles used for other purposes, such as injecting steroids, ear piercing, or tattooing. If you are interested in having your ears pierced or getting a tattoo, insist upon seeing a qualified person who uses either brand-new or properly sterilized equipment. Ask questions about the safety measures that are followed before undergoing any such procedure.

## Transmission of HIV from Mother to Offspring

HIV may be transmitted from mother to fetus during pregnancy or from mother to child through childbirth or breast-feeding.[24] Some 1,500 to 2,000 newborns in the United States are believed to be infected with HIV each year.[25] By the early 1990s in New York City, AIDS had become the leading cause of death of Hispanic-American children 1 to 4 years of age, and the second leading killer of African-American children in that age range.[26]

Present research suggests that mother-to-fetus transmission is most likely to occur during the birth process itself,[27] generally through exposure to infected blood and mucus in the birth canal. Babies of HIV-infected mothers stand about a one in three chance of being born infected with HIV.[28] It is possible that the risk of transmission during birth may be lessened by bathing the birth canal with antiviral agents prior to delivery and performing C-sections (Caesarean sections) prior to the rupture of the mother's membranes.

A genuine breakthrough in the war against AIDS occurred when researchers found that the antiviral drug *zidovudine* (AZT) administered to HIV-infected pregnant women reduced the rate of HIV infection in their newborns by two-thirds.[29] Scientists had suspected that AZT might help prevent AIDS in children of infected mothers by reducing the amount of the virus in the mother's bloodstream. The research focused on 477 pregnant women in the United States and France, all of whom were infected with HIV but were still healthy. Half took AZT during pregnancy and through labor and delivery, while the other half were given a placebo (a chemically inert pill). After birth, the babies were continued on the treatment the mothers had received for a six week period. Only 8% of the babies born to the AZT-treated women became infected with HIV, as compared to 25% of the babies born to women in the placebo group. Mild anemia was the only apparent side effect in the children. These striking results offer the best hope yet that many if not most cases of AIDS in children can be prevented by the use of already available drugs.

## Transmission of HIV through Blood Supplies

Prior to 1985, persons needing transfusions, such as hemophiliacs and surgical patients, faced an undetermined risk of infection from contaminated blood supplies. Over half of the hemophiliacs in the U.S. were infected with HIV in the early 1980s.[30] In 1985, the test to detect HIV antibodies (revealing the presence of HIV infection) became available, and blood banks began universal screening of donor blood. No U.S. hemophiliac is known to have contracted HIV from a blood transfusion since 1987, however.[31] Arthur Ashe, who died of complications from AIDS in 1993, reported that he had contracted HIV from a blood transfusion he received during a surgical operation performed before blood screening for HIV was introduced.[32]

Blood screening is not yet fool-proof, however. While transmission of HIV in the general population from blood transfusions have become quite rare since routine screening of blood supplies began, a remote possibility of infection continues to exist, with estimates suggesting a risk of infection of about 1 in 75,000 from receiving a single transfusion.[33] Because of this risk, however small it may be, many people today who are planning to have surgery store their own blood beforehand, so that they can be assured of a safe transfusion should it be needed.

HIV may also be spread by donor semen, such as that used in artificial insemination. Many sperm banks do not test donors for STDs. Cases have been reported of women who have become infected with hepatitis B, gonorrhea, trichomoniasis ("trich"), chlamydia, and even HIV via insemination with donor semen.

---

## HEALTH TIPS

▶ While the chances today of contracting HIV through a blood transfusion are remote (perhaps one in 75,000), you may wish to store your own blood in reserve in advance of a scheduled operation, so as to reduce the chances of contracting HIV to zero.

---

## Transmission from Health Care Workers to Patients

In 1991, the tragic case of young Kimberly Bergalis captured headlines as well as the nation's sympathy. Kimberly, who succumbed to AIDS in late 1991, was believed to have contracted HIV from her dentist during a routine office visit. Kimberly appeared before Congress shortly before her death and pleaded with lawmakers to enact legislation that would require HIV testing of health care workers. The dentist who had treated Kimberly died of AIDS after practicing for several years after learning that he was infected with HIV.[34] The dentist apparently infected several of his patients in addition to Kimberly. Although the mode of infection remains unknown, the most likely explanation is that the infection was spread from the dentist to his patients, perhaps through bleeding from an open cut on his hands, rather than from using contaminated instruments that were not properly sterilized.[35]

Although Bergalis's death was tragic, many health officials believe that the Florida case was an anomaly. The case of the dentist who treated Kimberly Bergalis is the only known case of HIV transmission from health care workers to patients in the U.S.[36] A case involving a French surgeon has also been reported.[37] Many health officials argue that regulations requiring massive testings of health-care workers are unnecessary and may unfairly affect the careers of health-care workers infected with the virus, as well as waste millions of dollars that might otherwise be spent on combating or preventing AIDS.[38] Still, a July 1991 *Newsweek* poll found that more than 90 percent of U.S. residents believe that health-care workers who are infected by HIV should be required to inform their patients of the infection.

### HIV Transmission from Patients to Health Care Workers

Health-care workers risk infection from accidental needle-sticks from needles used with infected patients. The Centers for Disease Control and Prevention (CD) reported that 42 people nationwide had become infected with HIV by the end of 1994 on the basis of on-the-job accidents. Another 91 were possibly infected on the job. Most of these cases involved lab technicians and nurses who accidentally pricked themselves with a needle or were cut by a scalpel that was used previously on an HIV-infected patients. Clearly, health providers need to exercise caution to avoid contact with infected blood.

# How HIV is Not Transmitted

There is much misinformation about the transmission of HIV. Let us consider some of the ways in which HIV is not transmitted:

1. *HIV is not transmitted from donating blood.* AIDS cannot be contracted by donating blood because needles are discarded after a single use. Unfortunately, many people have avoided donating blood because of unfounded fears of HIV transmission.

2. *HIV is not transmitted through casual, everyday contact.* There is no evidence of transmission of HIV through various kinds of casual contact. These include hugging someone or shaking hands, and bumping into strangers on buses and trains; handling money, doorknobs, or other objects that had been touched by infected people; sharing drinking fountains, public telephones, public toilets, or swimming pools; or by trying on clothing that had been worn by an infected person.[39] Nor is HIV known to be transmitted by contact with urine, feces, sputum, sweat, tears, or nasal secretions, unless blood is clearly visible in these fluids.[40] (Still, should you need to clean-up someone's urine, feces, nasal secretions, and especially blood, it would be wise to use rubber gloves and wash your hands thoroughly immediately afterwards).

3. *HIV is not transmitted by insect bites.* HIV is not transmitted by mosquito bites or from bites by other insects such as bedbugs, lice, or flies.[41] Nor can you get HIV from contact with animals.

4. *HIV is not transmitted by airborne germs or contact with contaminated food.* People do not contract HIV from contact with airborne germs, as by sneezing or coughing, or by contact with contaminated food or eating food prepared by a person infected with HIV.[42] (However, other disease-causing organisms, such as the virus that causes hepatitis A, may be transmitted by contact with contaminated food).

5. *HIV is not transmitted through sharing work or home environments.* HIV has not been shown to be transmitted from infected people to family members or others with whom they live through any form of casual contact, such as hugging or touching, or through sharing bathrooms, food or eating utensils, so long as there is no exchange of blood or genital secretions.[43] Nor have any cases of HIV transmission been documented based on nonsexual contact in schools or in the workplace.[44] There are some isolated reports of nonsexual transmission of HIV between two children living together in the same household, but investigators suspect that the route of transmission in these rare cases involved blood contact, involving perhaps the sharing of razor blades in one of the two reported cases thusfar and possible sharing of a toothbrush in the other (the infected child in this case had bleeding gums).[45]

AIDS-prevention programs are apparently raising public awareness about AIDS.[46] Evidence shows that young people are becoming better informed about the means by which AIDS is transmitted. For example, a 1988 national random sample of 15- to 19-year-old males showed a high level of knowledge about AIDS transmission.[47] Virtually all of the young men polled knew that AIDS could be transmitted by sharing hypodermic needles and by homosexual and heterosexual intercourse. Knowledge about AIDS is variable, however. Many young people sampled in a national survey of male adolescents and in another survey of male and female high school students in nine states[48] incorrectly believed that HIV can be acquired from donating blood or from a mosquito or insect bite.

---

## HEALTH TIPS

*YOU CANNOT CONTRACT HIV THROUGH:*

▶ Casual contact, such as shaking hands or touching an infected person.

▶ Mosquito or other insect bites or contact with animals.

▶ Sharing a domicile or work or school environment with an infected person.

▶ Breathing airborne germs.

▶ Eating contaminated food (but ingesting contaminated food may transmit the virus causing hepatitis A—see Chapter 11).

---

Unfortunately, knowledge of AIDS transmission is not sufficient to stem the epidemic. Researchers find that information about preventing AIDS often does not translate into safer sex practices.[49] Despite efforts to educate the public about the dangers of unprotected sex, studies show continued low rates of use of condoms among two key population groups: sexually active teens and adults with multiple sex partners.[50] While awareness of the prospects of HIV infection and AIDS is often insufficient to motivate people to engage in "safer sex," it is an indispensable step, however. Later in the book you will find specific suggestions for translating awareness of HIV infection into a course of action to reduce your risk of infection.

---

# A Closer Look

## Are You at Risk of Infection?

Several risk factors for HIV have been identified since HIV first appeared in the U.S. in 1978. If you answer "Yes" to any of the following questions, you may be at increased risk of infection. Since 1978...

▶ Have you engaged in needle sharing in injecting drugs or steroids?

▶ Have you, if you are a male, engaged in sex with other men?

▶ Have you had sex with someone you suspected was infected with

HIV?
- ▸ Have you had a sexually transmitted disease (STD)?
- ▸ Did you receive a blood transfusion or blood products sometime between 1978 and 1985?
- ▸ Have you had unprotected sex with anyone who would answer yes to any of the above questions?

The CDC recommends that anyone who answers yes to any of these questions should consult a trained counselor to discuss the need for an HIV test.[51] Consultation is even more important for women who answer yes to any of the above questions who may be planning to become pregnant.

People who are also at greater risk of infection include those who do not know the risk behavior of their sexual partners or have had many sexual partners during the past 10 years. All in all, the only way to know whether you are infected with HIV is to have an HIV-antibody blood test. However, HIV testing raises important emotional and lifestyle concerns (see Chapter 5) and should be undertaken only if you have carefully considered what it would mean to receive a *positive* (HIV-infected) result. According to the CDC, people considering HIV testing should discuss the meaning of the test with a qualified health professional *both before and after* the test is conducted.[52]

Let us end this chapter on a positive note. Given the lethality of HIV, it is indeed fortunate that transmission of the virus requires intimate sexual contact or exchange of blood and does not occur through casual contact. Because the modes of transmission are so limited, people can greatly reduce the chances of contracting the virus by following the guidelines for prevention discussed later in the book.

## CHAPTER FOUR

# WHAT YOU SHOULD KNOW ABOUT WOMEN AND AIDS

*DID YOU KNOW THAT?*

▸ *Though the risks of HIV infection and AIDS fall disproportionately in the U.S. on poor women from inner-city urban areas, HIV and AIDS cut across all racial, social and economic boundaries.*
▸ *Women who abuse cocaine are at increased risk of HIV infection.*
▸ *AIDS may become expressed in different ways in women than in men.*
▸ *Women with AIDS die sooner than do men with AIDS.*

*Lily was not supposed to get AIDS. She was heiress to a cosmetics fortune. She had received her bachelor's degree from Wellesley and had been enrolled in a graduate program in art history when she came down with intractable flulike symptoms and was eventually diagnosed as having AIDS.*

*"No one believed it," she said. "I was never a male homosexual in San Francisco. I never shot up crack in the alleys of The Bronx. My boyfriends didn't shoot up either. There was just Matthew . . ." Now Lily was 24. At 17, in her senior year in high school, she had had a brief*

*affair with Matthew. Later she learned that Matthew was bisexual. Five years ago, Matthew died from AIDS.*

*"I haven't exactly been a whore," Lily said ironically. "You can count my boyfriends on the fingers of one hand. None of them caught it from me; I guess I was just lucky." Her face twisted in anger. "You may think this is awful," she said, "but there are times when I wish Jerry and Russ had gotten it from me. Why should they get off?"*

*Lily's family was fully supportive, emotionally and, of course, financially. Lily had been to fine clinics. Physicians from Europe had been brought in. She was on a regimen of three medicines: two antiviral drugs, which singly and in combination had shown some ability to slow the progress of AIDS, and an antibiotic intended to prevent bacterial infections from taking hold. She took some vitamins--not megavitamin therapy. She exercised almost daily when she felt up to it, and she was doing reasonably well. In fact, there were times when she thought she might get over her illness.*

*"Sometimes I find myself thinking about children or grandchildren. Or sometimes I find myself looking at all these old pictures [of grandparents and other relatives] and thinking that I'll have silver in my hair, too. Sometimes I really think this is the day the doctors will call me about the new wonder drug that's been discovered in France or Germany."*

*"I want to tell you about Russ," she said once. "After we found out about me, he went for testing, and he was clear [of antibodies indicative of infection by the AIDS virus]. He stayed with me, you know. When I wanted to do it, we used condoms. A couple of months later, he went for a second test and he was still clear. Then maybe he had second thoughts, because he became impotent--with me. We'd try, but he couldn't do anything. Still he stayed with me, but I felt us drifting apart. After a while, he was just doing the right thing by staying with me, and I'll be damned if anyone is going to be with me because he's doing the right thing."*

*Lily looked the [interviewer] directly in the eye. "What sane man wants to play Russian roulette with AIDS for the sake of looking like a caring person? And I'll tell you why I eventually sent him away," she added, tears welling, "the one thing I've learned is that you die alone. I don't even feel that close to my parents anymore. Everyone loves you and wishes they could trade places with you, but they can't. You're suddenly older than everyone around you and you're going to go alone. I can't tell*

*you how many times I thought about killing myself, just so that I could be*
*the one who determines exactly where and when I die--how I would be*
*dressed and how I would feel on the final day. "*
*Lily died in 1992.[1]*

Not so long ago, there were a number of erroneous assumptions about HIV and AIDS—assumptions that have had a disproportionate negative impact on women. They included the notions that women were at low risk of HIV infection and that it was difficult to be infected via heterosexual intercourse. It was also assumed that HIV infection and AIDS would follow the same course in men and women—a belief that may be connected with delayed diagnosis and intervention with women. Let us separate fact from fiction. Here's what you should know about women and AIDS:

# HIV Infection and AIDS among Women is Increasing at Alarming Rates

Although AIDS was once considered a syndrome that afflicted only gay men, the numbers of U.S. women with AIDS are on the rise. By the mid-1990s, the number of women with AIDS in the United States was doubling every year.[2] By 1996, women accounted for 20% of the new cases of AIDS.[3] Worldwide, women account for about one-third of AIDS cases.[4] In some parts of Africa, such as Uganda, women now account for the majority of adult AIDS cases. If present trends continue, most new cases of HIV worldwide by the year 2000 will be among women.[5]

With increasing numbers of AIDS cases in women reported in recent years, it is not surprising that the number of deaths due to AIDS among women is on the rise. By the mid-1990s, AIDS had become the fourth leading killer of women in the 25 to 44 age range.

## Reported AIDS Cases among Women May Represent the Tip of the Iceberg

As noted in Chapter 2, AIDS is increasing more rapidly in the U.S. among women than among men. The numbers of reported AIDS cases among women may represent the tip of iceberg, because symptoms may not develop for years after infection with HIV.[6] A leading researcher on AIDS, Alexandra Levine of the University of Southern California Medical Center, noted the following:[7]

> We have not yet begun to see what's going to happen with women and HIV. We are now with women in the same situation as we were for gay men in 1983 or 1984. It can happen to you or to me or to any of us. This is a sexually transmitted disease. Period. You must think of yourself as potentially at risk. It's the only way we're going to get on top of this epidemic.

## Though AIDS has Struck Economically Disadvantaged Women Most Heavily, Anyone Can Contract the AIDS Virus and Develop AIDS

Among women in our society, the risks of HIV infection and deaths from AIDS falls most heavily on the least advantaged: poor women, mostly African-American or Hispanic, who live in depressed urban areas.[8] Still, as the case of Lily indicates, HIV infection and AIDS cut across all racial, social, and economic boundaries.

Overall, African-American and Hispanic women account for about three of four cases of women with AIDS in the United States, although they constitute only 19 percent of the female population.[9] White women account for about one-quarter of the cases of AIDS among women.

The death rates among women with AIDS also fall most heavily on women of color. African-American women are nine times more likely to die from AIDS than white women.[10] In New York and New Jersey, AIDS has become the leading cause of death among African-American women aged 14 to 44.[11]

But HIV infection and AIDS cut across all racial and economic groupings. Anyone—man, woman or child—can contract HIV. The virus doesn't care about your nationality, skin color, or ethnic class. It doesn't know whether you are black, white, or brown or whether you attend religious services regularly or always brush your teeth. All it "knows" is how to replicate itself and disable your immune system once it finds a port of entry into your body.

Yet, many women remain unaware that they are at risk of contracting

AIDS.[12]  Many HIV-infected women go undiagnosed until they develop AIDS or give birth to an HIV-infected baby.

# HIV Transmission in Women

In the early years of the epidemic, most women in the United States with AIDS had contracted the disease through intravenous drug use. By 1992, heterosexual sex had surpassed injecting drug use as the primary means of transmission of HIV in women.[13]  Though drug abuse of the male partner is implicated in a majority of heterosexually transmitted AIDS cases in women, some women have contracted HIV from heterosexual men who do not abuse drugs. We do not hear too much about them, however. Women too, as noted in Chapter 3, stand a greater chance of contracting HIV through heterosexual sex than do men.

---

## HEALTH TIPS

▶  Any woman who engages in unprotected sex with partners who are not known to be uninfected should consider herself at risk of HIV infection, not just women who are IDUs or have partners who are bisexual or IDUs. AIDS is not a disease of any one group or gender.

---

## Cocaine Use and Risk of HIV Infection

Rates of HIV infection are higher among women who use cocaine, perhaps because they often engage in high-risk sexual behaviors, such as prostitution or unprotected intercourse with IDUs, not because of cocaine use per se. Since injectable drug use and cocaine use is more prevalent among women in economically distressed urban areas, it is not surprising that HIV infection has spread most rapidly among these groups.

# Does HIV Infection and AIDS Follow a Different Course in Men and Women?

Many questions remain about how HIV infection and AIDS affect women. Much of what we know about how HIV infection and AIDS affect the body is based on research conducted on gay men. We do not know whether HIV infection and AIDS follow the same course in women. AIDS may also have symptoms that go unrecognized or are misdiagnosed in women, leading to delayed diagnosis and intervention.[14] Moreover, much of the knowledge gained about the effects of AIDS drugs has also been drawn from research on treatment of men, so questions remain about the effectiveness of these drugs on women and children.

AIDS activists claimed that the criteria for diagnosing AIDS failed to recognize indicator diseases that may be linked to HIV infection and AIDS in women—such as pelvic inflammatory disease (PID), precancerous cervical disease, and vaginal candidiasis (yeast infection). (The great—great! majority of yeast infections are unrelated to HIV, however). The appearance of different sets of symptoms in HIV-infected women is one of the reasons that the government revised the diagnostic criteria for AIDS by adding certain indicator diseases that affect women and expanding the diagnosis to include measurable damage to the immune system (CD4 cell counts of less than 200).

While many questions about the course of HIV infection and AIDS in women remain, evidence shows that AIDS kills women more quickly than men.[15] Delayed diagnosis and treatment of women with HIV and AIDS may play a role in explaining why women with AIDS tend to die sooner.[16] The fact that AIDS afflicts poor women disproportionately may partly explain delays in receiving appropriate medical care, including access to the new "AIDS cocktail" of antiviral drugs, since disadvantaged women generally have poorer access to medical care.[17]

# Women's Prevention Choices More Limited

Though AIDS prevention programs have focused on increasing condom use among sexually active people, many women cannot compel their partners to use condoms.[18] Moreover, the one barrier method available to women, the female condom, has not been established to provide effective protection against STDs. Sexual counseling programs stress that men need to take more responsibility for wearing condoms and that both partners should learn how to talk about HIV prevention.

## CHAPTER FIVE

# WHAT YOU SHOULD KNOW ABOUT HIV TESTING AND TREATMENT OF HIV/AIDS

*DID YOU KNOW THAT?*

- *HIV infection is not diagnosed by detecting the presence of the virus in the blood system. How, then, is it diagnosed?*
- *You can get a negative (non-infected) HIV test result and still be infected by HIV.*
- *Your chances of falsely testing positive for HIV are about 1 in 100,000 on the basis of a combination of presently available tests.*
- *Many scientists are hopeful that a vaccine for AIDS will eventually be found.*
- *The best approach to stemming the spread of AIDS is prevention.*

How would you know if you were infected with HIV or had AIDS? How would you know if your partner were infected? Tests for detecting HIV infection have been available since the mid-1980s. A simple blood test can determine whether a person is infected with HIV by means of detecting whether specific antibodies to HIV are present in the person's blood system. Determining whether a person has

AIDS is a more complicated matter, and depends upon the presence of certain diagnostic indicator diseases or an abnormally low CD4 cell count in people with HIV.

Testing for HIV infection is widely available and relatively straight-forward. There are even home tests now available. Yet HIV testing is no cut-and-dried matter, however, as suggested by the following comments:[1]

> PHIL, 20: The only change I've made as a result of the AIDS threat is to use condoms as protection. But when I find a girl-friend who is really special, I plan to be tested for AIDS. Hopefully this would alleviate any fears she might have, and it would show her how much I really care.

> MELANIE, 18: How to bring up AIDS with someone you'd like to sleep with confuses me. I mean, what do you say? "Gee, honey, I love you and want to have sex with you, but can you please take an AIDS test today and then lock yourself up for six months so I'm sure you won't sleep with anyone else and then take the test again? Then maybe I'll sleep with you."

> DON, 25: I "came out" just as AIDS hit the media in metro-Boston as the Gay Plague. During the first year or so I did very little to practice safe sex. Now, I will not do anything that is against the safe-sex guidelines. I often fear the days I did not follow these guidelines. I often think of being tested, although I am scared. I would not be able to have sexual relations or even date should I test positive, because I am too moral to pretend that nothing is wrong, and too uncomfortable coming forward with this information to my partners. Sometimes I think it is best not to know and to always behave responsibly; other times I think I should know so that I can make plans for my life and if I test negative, seek a long-term relationship "armed" with this information.

Concerns and motivations regarding HIV testing can be complex and contradictory. While all too many people, even in the age of AIDS, engage in sexual practices without regard to the potential risk, many have become more cautious in their sexual practices. This vigilance represents a broad spectrum of response. Individuals may feel that they need to know more about one another before they proceed with a sexual relationship. Some will not even consider such a relationship in the absence of an HIV test that shows a prospective partner to be clear of HIV infection. Others might even take the initiative to have themselves tested, as a way of assuring a partner that they themselves are safe, or, conversely, to alert their partners that they are indeed infected with HIV. As logical as these measures may seem when weighed against the potential risks, concerned partners may foresee emotional ramifications to such a direct approach. How can one broach such a subject at all? How can a personal relationship get off the ground with concerns and suspicions at its very start? How do you suggest to someone with whom you would like a sexual relationship that he or she should be tested for HIV?

Yes, but what about obligation? Many believe that people at risk of HIV infection who have not taken steps to determine their HIV status before entering sexual relationships have acted immorally, irresponsibly, and even criminally. A prospective partner has the right, it is argued, to have such information. Its disclosure can lead to appropriate precautions including safer sex practices or the avoidance of needle-sharing.

Once again, however, the issues become muddied. Some who are opposed to widespread testing argue that testing is unnecessary and alarmist for people who have not engaged in high-risk sexual or injection practices. Opponents further point out that a person whose HIV-positive status becomes known becomes "branded." He or she may fall victim to harsh discrimination regarding employment, housing, or medical and life insurance. Moreover, opponents argue that safer sex guidelines should be practiced as a matter of course in this age of AIDS, regardless of HIV status. Finally, they voice concern that knowledge that one carries HIV antibodies, or has AIDS, can be emotionally devastating and that people may be ill-prepared to deal with the enormous stress experienced in the wake of this realization. On the other hand, what they don't know may hurt them, because early detection and intervention with HIV-fighting drugs may prolong health and even save lives of infected people. Knowing that one is HIV-infected may also encourage one to take better care of one's health, avoid stress, and adopt a balanced diet, which may buttress one's immune system.

---

**HEALTH TIPS**

▸ HIV testing raises important emotional, medical, ethical, moral, and lifestyle concerns. You should seriously weigh these concerns in deciding whether you should undergo testing or insist that your partner be tested.

---

Home testing for HIV is highly reliable and protects the user's anonymity.[2] The user draws a small sample of blood and mails in a coded sample to a laboratory for analysis. However, concerns have been voiced that people may not receive counseling before and after testing. Companies promoting home testing are offering telephone counseling services. Whether these services will be used, and whether they will be helpful, remain open questions.

In sum, the issues of whether to be tested or to insist that one's partner be tested raises important concerns. Acquiring information about HIV testing and AIDS diagnosis may help you make a better informed decision. But you must decide for yourself whether you are psychologically prepared to undergo HIV testing. *However, a person who experiences unexplained symptoms and has reason to suspect that he or she may be infected with HIV should (no, must!) seek a medical consultation immediately.* Here's what else you should know about HIV testing and AIDS diagnosis:

# HIV Testing

The most widely used test for HIV infection is the enzyme-linked immunosorbent assay (ELISA, for short). ELISA does not directly detect HIV in the circulatory system. Instead, it reveals HIV antibodies. People may show an antibody response to HIV long before they develop symptoms of infection. Drawing the blood takes only a few minutes and the results are usually available in two or three weeks. Although the antibody test does not directly reveal the presence of the virus, a positive ("seropositive") test result generally indicates that the person has been infected with HIV. (We say generally, rather than always, because fetuses may receive antibodies from infected mothers, but not the virus itself. However, some fetuses do become infected with the virus). A negative ("seronegative") outcome means that antibodies to HIV were not detected.

Antibodies to HIV may show up in the blood system long before any symptoms of the infection appear. So simply because you or your partner look and feel well does not guarantee that either of you are free of HIV. Yet, *antibodies to*

*HIV may not show up for perhaps three to six months or even longer after initial exposure.* So a misleading negative test result may occur in a person infected with HIV who does not yet produce HIV antibodies. This means that an initial negative test may need to be confirmed by a later retesting.

---

### HEALTH TIPS

▶  A lack of symptoms does not mean that you are free of HIV infection. You may show a positive test result even if you look and feel perfectly fine.

▶  Since HIV antibodies may not show up in the blood for months after exposure, only repeated testing can confirm an initial negative result.

---

ELISA fails to detect HIV antibodies in about three in one thousand cases (a *false* negative result). In about one in one hundred cases, it produces a false positive result (an incorrect seropositive finding).[3] When people receive positive results on the enzyme-linked immunosorbent assay, the Western blot test can be performed to confirm the findings.[4] The Western blot test detects a particular pattern of protein bands that are linked to the virus. The combination of ELISA and the Western blot test reduces the chances of a false seropositive finding to about 1 in 100,000.[5]

Blood tests for HIV detect markers of HIV infection, such as the presence of HIV antibodies (in the case of ELISA) and certain protein formations (in the case of the Western blot test) associated with the presence of the infection. They do not directly test for the presence of the virus. Efforts to develop more precise tests that directly detect the AIDS virus are underway and may be available by the time this book reaches your hands.

# A CLOSER LOOK

## Questions to Ask Yourself Before Deciding to Have an HIV Test (Or Before Asking Your Partner to Have One)[6]

Deciding whether you should be tested, or asking your partner to be tested, raises important issues. First, how likely is it that you fall into a high-risk category? Have you, or anyone with whom you've had sexual relations, engaged in risky sexual behavior or injection practices since the start of the epidemic, or received a blood transfusion between 1978 and 1985? Have you had sex with anyone whom you suspect might have been infected with HIV? (Would your partner answer yes to these questions?) If so, have you (and your partner) thought through the pros and cons of being tested?

How prepared would you be to tell your sexual partners if you tested positive—not only your current partners but any partners you've had since the epidemic began? Do you recognize any ethical and moral responsibility to inform your present and past partners?

You also need to consider your psychological preparedness for testing. How prepared are you to deal with the emotional trauma of receiving a positive test result? How prepared is your partner? Have you thought through a strategy for coping with a positive test result, such as drawing upon support from a network of friends and professionals to help you cope with the possible trauma of learning that you are infected?

Have you considered what effects a positive result would have on your primary relationship? Have you considered how your partner might react? Are you prepared for the changes that may occur in the relationship, or for the possibility that the relationship may not survive? On the other hand, do you feel a sense of ethical or moral responsibility to your sexual partners to check out your HIV status?

Have you considered how a positive result would affect other aspects of your life, such as your employment, career plans, family relationships, and so on? Would you be prepared to handle the social stigma attached to people infected with the AIDS virus? Society has been short on tolerance for people infected with the AIDS virus. Societal prejudices and patterns of discrimination strain even those who are the best able to cope with living with HIV. How might you counter the discrimination and prejudice that you are likely to face from others?

One alternative to consider is *anonymous* testing. Through anonymous testing, your test results will not be identified with your name or revealed to anyone but you. To find out how to obtain an anonymous HIV test in your area, you can contact the Centers for Disease Control National AIDS Hotline at 1-800-342-AIDS. You need not give your name. Home testing also allows you to maintain your anonymity.

You might also wish to discuss with your doctor the option of confidential testing. In confidential testing, only your doctor is given your test results, which are then kept in your doctor's records. Your doctor has an obligation to keep your medical records confidential. However, you may need to sign a release to have your records disclosed to others in some circumstances, such as when applying for life insurance or should you require care in a hospital. Discuss with your doctor any questions you have regarding the issue of confidentiality before deciding to go ahead with an HIV test.

# Progression of HIV Infection to AIDS

A seropositive test result means that HIV antibodies have been found, but does not indicate if or when the individual will develop a full-blown case of AIDS. We can't say for certain that a person who tests positive for HIV will go on to develop AIDS. Many people with HIV today are living ten, fifteen, or more years without developing AIDS.

# Treatment of HIV Infection and AIDS

Despite promising developments in the treatment of HIV infection and AIDS, there is neither a cure for AIDS nor a vaccine to protect people from the disease. Doctors have been able to slow the progression of the illness and in some cases reduce the presence of HIV in the blood to undetectable levels. Doctors may also be able to effectively treat some of the opportunistic infections that take hold in AIDS patients whose weakened immune systems leave them vulnerable to a host of diseases.

## Treatment of People with HIV and AIDS

HIV infection and AIDS are generally treated through a combination of health counseling, emotional support, and use of antiviral drugs. People with HIV and AIDS may be counseled to help them avoid stress, to manage unavoidable stress more effectively, and to adopt more healthful behaviors, such as following a proper diet and getting sufficient sleep. Avoiding alcohol and illicit drugs, and maintaining a regular exercise regimen, may also help strengthen the immune system and decrease the chances of opportunistic infections taking hold. People with HIV and AIDS are also counseled about the importance of practicing safer sex techniques if they continue to be sexually active.

Emotional support is an important feature of HIV/AIDS treatment. The majority of people with HIV and AIDS encounter serious adjustment problems. Psychological intervention may help HIV and AIDS patients maintain a sense of hope and find a purpose in life, as well as help relieve emotional distress. Self-help groups, support groups, and organized therapy groups have provided assistance to people with HIV and AIDS, their families, and friends.

Training in stress-management techniques, such as relaxation techniques and positive mental imagery, may help people with HIV and AIDS improve their ability to handle stress. Social support from friends and family also may help to buffer the effects of stress. However, people with HIV and AIDS often suffer social rejection and find their social relationships to be more stress-inducing than stress-reducing.

## Development of a Vaccine: Still Years Away

HIV can mutate into forms that are resistant to particular antiviral drugs or vaccines. The genes making up HIV mutate a million times more rapidly than human genes, leading scientists on wild goose chases as they try to eradicate a killer that keeps changing its form so often. A vaccine that might protect against one strain, or a drug that could render a death blow, may hold no value against a mutated variant.[7]

Despite these challenges, many scientists believe that a safe and effective vaccine will eventually be developed. A number of experimental vaccines are now being tested on uninfected people. Still others are in the stage of animal experimentation. With the number of studies required before a vaccine can be brought to market, none are expected to be available to the public for many years. Several vaccines have proved somewhat effective in animal trials. Yet AIDS experts caution that the development of an ideal vaccine against AIDS—one that would be safe, inexpensive, and render lifetime protection against all strains of the disease with a single dose—is perhaps decades away at the earliest.[8] In the meantime the best (make that only) form of protection you have against HIV or AIDS is to take steps to reduce your chances of contracting the virus.

## Drug Treatment: Suddenly, Hope

The most widely used AIDS drug, AZT (*zidovudine*), may slow the progression of HIV infection from the asymptomatic to the symptomatic state.[9] However, the drug eventually loses its effectiveness as more resistant forms of the virus begin to flourish. AZT may also produce serious side-effects, such as suppression of bone marrow function, which leads to anemia and lowers the white blood count, further reducing the body's ability to combat infections.[10]

AZT belongs to a class of drugs called *reverse transcriptase inhibitors* because they block replication (reproduction) of HIV by inhibiting an enzyme, *reverse transcriptase*, that the virus needs to reproduce. Other types of reverse transcriptase inhibitors include ddI, ddC, d4T, and *3TC*. Combining AZT with these similar drugs keeps the pressure on HIV, delaying the progress of HIV infection to AIDS over AZT alone.[11]

Most promising yet is a new generation of drugs, the *protease inhibitors*. *Saquinavir* and *ritonavir* are two examples. These drugs take aim at a different phase of the life cycle of HIV by interfering with the protease enzyme which HIV needs to reproduce itself.[12] Now, a combination of antiviral drugs—the so-called AIDS cocktail—consisting of reverse transcriptase inhibitors, such as AZT and 3TC, along with a protease inhibitor, has produced dramatic results in many cases, reducing HIV in the blood stream to undetectable levels.[13] Scientists caution that even though HIV may not be detected in the blood it may still be lurking in other body tissue. Only time will tell whether the remarkable progress seen in many patients treated with the "AIDS cocktail" will be maintained over time. Perhaps other drugs now on the planning table will be needed to rout HIV if it returns in force.

The advent of these new drug combinations has fueled hope that AIDS may become a manageable disease, as opposed to terminal illness.[14] Some respected scientists have even expressed hopes of a *possible* cure.[15]

Already we are seeing a significant impact of this new approach to treatment. The number of AIDS deaths nationwide dropped 13% in the first half of 1996, as compared to the same period in 1995, the first drop since the epidemic began in 1981. The Centers for Disease Control and Prevention (CDC) credits the drop to the advent of new drug therapies and increased access to treatment.[16] Health officials in New York City, the city hardest hit by the AIDS epidemic, reported a 30% drop in deaths in 1996 due to AIDS.[17] A doctor in Georgia who treats AIDS patients commented that a few years ago he would see people die from AIDS in a few months. Now, he sees them leaving the hospital and going back to work.[18]

Still, AIDS continues to claim more than 44,000 lives annually in the U.S.

The new drug therapy also has its limitations. For one thing, it is very expensive, averaging between $10,000 and $15,000 per year.[19] Many people who might benefit from the drug regimen may not receive it because they lack the financial resources or health insurance coverage needed to foot the bill. Moreover, we don't know whether the AIDS cocktails will maintain their effectiveness over time.

Despite advances in treatment, don't place your bets on the development of a quick cure or a surefire vaccine. The best defense against HIV/AIDS is prevention. For ways of protecting yourself, see Chapter 13.

# STD FACT SHEET: ACQUIRED IMMUNODEFICIENCY SYNDROME (AIDS)

| | |
|---|---|
| **WHAT IT IS** | A viral syndrome in which the body's immune system becomes stripped of its ability to defend itself against life-threatening diseases |
| **WHAT CAUSES IT** | The *human immunodeficiency virus* (HIV) |
| **HOW IT IS TRANSMITTED** | Sexual contact involving vaginal or anal intercourse, or oral-genital contact with an infected partner; sharing of contaminated hypodermic needles; transfusion with contaminated blood supplies; from an infected mother to fetus in-uteru or to a newborn during delivery |
| **SIGNS AND SYMPTOMS** | May be asymptomatic at first or produce temporary flu-like symptoms shortly following initial infection. Person may remain asymptomatic for years or progress to a symptomatic stage of HIV infection, followed by progression to full-blown AIDS which is characterized by fever, unexplained weight loss, night sweats, fatigue, and the appearance of opportunistic illnesses like Kaposi's sarcoma and PCP. |
| **HOW IT IS DIAGNOSED** | HIV infection is diagnosed by blood tests that detect HIV antibodies in the blood. AIDS is diagnosed by the appearance of indicator (opportunistic) diseases or a fall-off in the number of CD4 cells to below 200 per cubic millimeter in persons infected with HIV. |
| **HOW IT IS TREATED** | While no cure is available, a combination of antiviral drugs including reverse transcriptase inhibitors and protease inhibitors has brought hope that AIDS may become a manageable disease, much like diabetes, rather than a terminal illness. Stress-management, and proper diet and exercise may help improve psychological and physical functioning. |

## CHAPTER SIX

# WHAT YOU SHOULD KNOW ABOUT GONORRHEA

## *DID YOU KNOW THAT?*

▸ *If you are 20 to 24 years of age, you are in the age group most at risk of contracting gonorrhea.*

▸ *About four of five women who contract gonorrhea experience no symptoms, but are capable of passing the infection along to their sexual partners.*

▸ *Gonorrhea can cause internal damage even in the absence of symptoms or if initial symptoms disappear.*

▸ *You can develop a gonorrheal infection of the throat by performing oral sex on an infected partner.*

▸ *The chances of contracting gonorrhea from just one sexual contact with an infected partner are about 50 percent for women and 20 to 25 percent for men.*

▸ *Untreated, gonorrhea can lead to infertility in men and women.*

▸ *Some strains of gonorrhea are resistant to penicillin.*

Has your doctor ever told you that you had strep throat? Scarlet fever? Pneumonia? Bronchitis? These are but a few of the diseases that are caused by tiny, one-celled organisms called **bacteria** (singular form, bacterium), which come in various forms and shapes. While we tend to think of bacteria as harmful organisms (indeed some are), most are actually helpful rather than harmful. Bacteria are the organisms responsible for disintegrating organic waste materials.

They also play vital roles in our digestive system. However, disease-causing bacteria are responsible for a host of diseases, including such STDs as gonorrhea, syphilis, chlamydia, and certain vaginal infections. This chapter focuses on gonorrhea.

## What is Gonorrhea?

Gonorrhea ("the clap" or "the drip") is a bacterial infection caused by the gonococcus bacterium. Gonorrhea has been known since ancient times. The term *gonorrhea* is credited to the Greek physician Galen, who lived in the second century A.D. Yet the bacterium responsible for the disease was not isolated until 1879, when it was first identified by Albert L.S. Neisser, after whom it was named *Neisseria gonorrhoeae*.

## How Widespread is Gonorrhea?

Gonorrhea is one of the most common types of sexually transmitted diseases in our society, accounting for an estimated 1.1. million cases annually in the U.S.[1] Since many other cases go unreported, the actual number of cases may be as high as several million a year.

Gonorrhea strikes predominantly among young people, typically between the ages of 20 and 24.[2] If you are sexually active, you should consider yourself at risk, irrespective of your age.

## What Are the Symptoms of Gonorrhea?

In men, symptoms usually appear within two to five days following infection. At first, a clear penile discharge appears, which, a day or so later turns yellow to yellow-green, thickens, and becomes puslike. Inflammation of the urethra (the tube that carries urine from the bladder out through the penis in men, or through the urinary opening in women) occurs, and urination is accompanied by a burning sensation. Thirty to 40 percent of males have swelling and tenderness in the lymph glands of the groin. Inflammation and other symptoms may become chronic if left untreated.

Gonorrhea in women primarily infects the cervix (the lower end of the uterus that adjoins the vagina), causing **cervicitis**, which is often accompanied by a yellowish puslike discharge that inflames the genitals. Women, like men, can have burning sensations when they urinate if the infection spreads to the urethra. Although such symptoms are possible, about four women in five have no noticeable symptoms (are asymptomatic). For this reason, they are unlikely to seek treatment unless the disease advances to produce more serious symptoms. They may also innocently infect another sex partner. Although infected men are more likely to have noticeable symptoms, they too may be infected without realizing it.

---

### HEALTH TIPS

▸ Most women and some men who become infected with gonorrhea experience no symptoms. With or without symptoms, the infection may cause internal damage and be transmitted to others.

---

Even without treatment, the initial symptoms of gonorrhea, if they do occur, often abate within a few weeks. Thus, victims may think of gonorrhea as being no worse than a bad cold. The truth is that even though the early symptoms may fade, the gonococcus bacterium may continue to cause damage within the body.

---

### HEALTH TIPS

▸ While the initial symptoms of gonorrhea may abate, if left untreated, the infection can spread within the reproductive system, causing damage to internal organs.

---

# How is Gonorrhea Transmitted?

The gonococcus bacterium that causes gonorrhea requires a warm, moist environment. The mucous membranes along the urinary tract in both men and women and along the cervix in women provide such an environment for the bacterium to flourish. Outside the body, the gonococcus bacterium dies in about a minute.

Despite what you may have heard, there is no evidence of transmission of gonorrhea from touching toilet seats or other dry objects.[3] However, gonorrhea may be contracted by contact with a moist and warm towel or sheet used immediately beforehand by an infected person.[4] Such occurrences are rare, however. In almost all cases, gonorrhea is transmitted sexually through vaginal or anal intercourse, oral-genital sex, or from mother to newborn during delivery.[5]

---

## HEALTH TIPS

▸ No, you will not contract gonorrhea by touching a toilet seat. However, some other STDs, such as "crabs," can be contracted from contact with toilet seats used previously by infected persons, as we shall see in later chapters.

---

Gonorrhea can also infect other parts of the body than the genitals. A person who performs oral sex (fellatio) on an infected man may contract a gonorrheal infection of the throat, called *pharyngeal gonorrhea*. Pharyngeal gonorrhea may also be transmitted, but less commonly so, by mouth-to-mouth kissing and by oral sex performed on an infected woman (cunnilingus). The eyes provide a good environment for the bacterium. Thus, a person whose hands come into contact with infected genitals and who inadvertently touches his or her eyes afterward may infect them. Infants have contracted gonorrhea of the eyes **(gonococcal ophthalmia neonatorum)** when passing through the birth canals of infected mothers. This disorder may cause blindness but has become rare because the eyes of newborns are treated routinely with antibiotic ointment toxic to gonococcal bacteria.

Gonorrhea of the pharynx and the rectum are fairly common among gay males and are spread by oral and anal intercourse, respectively. A gonorrheal infection of the cervix in a woman may be transferred to her rectum if the couple engages in anal intercourse directly following vaginal intercourse. Gonorrhea is less likely to be spread by vaginal than penile discharges. Thus, lesbians are less likely than male homosexuals to contract the disease. Still, any intimate sexual contact with an infected partner may transmit the disease.

<table>
<tr><td colspan="2" align="center">HEALTH TIPS</td></tr>
<tr><td>▸</td><td>Gonorrheal infections may be spread from one part of the body to another, such as from the cervix to the rectum in the case of an infected woman who engages in anal intercourse directly following vaginal intercourse.</td></tr>
</table>

# What Are the Risks of Contracting Gonorrhea?

Gonorrhea is highly contagious. Women stand a slightly greater than 50 percent chance, and men a 20 to 25 percent chance, of contracting gonorrhea after just one sexual exposure to an infected partner.[6] The risks to women are believed to be greater because women retain infected semen in the vagina following intercourse. The risk of infection increases with repeated exposure. In men, for example, the risk of infection climbs to 60 to 80 percent following four sexual encounters with one infected women, or one sexual encounter with each of four infected women.[7]

# What Are the Risks Associated with Untreated Gonorrhea?

When gonorrhea is not treated early, it may spread within the body, striking the internal reproductive organs. In men, it may result in *epididymitis* (an infection of the epididymis, the coiled tube in the back of each testicle through which sperm pass as they mature), which can cause fertility problems. Occasionally the kidneys are affected. The bacterium may spread through the cervix in women to the uterus, **fallopian tubes**, ovaries, and other parts of the abdominal cavity, causing **pelvic inflammatory disease** (PID).

Symptoms of PID include cramps, other forms of abdominal pain and tenderness, cervical tenderness and discharge, irregular menstrual cycles, coital pain, fever, nausea and vomiting. PID may also be asymptomatic. Whether or not women experience symptoms, PID can cause scarring that blocks the fallopian tubes, which can lead to infertility and ectopic (tubal) pregnancies. More than 100,000 women in the U.S. become infertile as the result of PID each year.[8] Ectopic pregnancies develop in the fallopian tube rather than the uterus. They must be surgically removed (aborted) to avoid rupturing the tube.

PID is a serious illness that requires aggressive treatment with antibiotics to fight the underlying infection.[9] Surgery may also be needed to remove the infected tissue. Unfortunately, many women only become aware of a gonococcal infection when they experience the discomfort of PID.

# How is Gonorrhea Diagnosed?

Diagnosis of gonorrhea involves clinical inspection of the genitals by a physician, followed by the culturing and examination under a microscope of a sample of genital discharge.[10] If you have any symptoms of burning urination, a yellowish, puslike discharge from the penis or vagina, or any feelings of pelvic pain or discomfort, consult your physician.

# How is Gonorrhea Treated?

Gonorrhea is treated effectively with antibiotics. While penicillin was once the treatment of choice, the rise of penicillin-resistant strains of *Neisseria gonorrhoeae* has led to the use of alternate antibiotics.[11] A single injection of the antibiotic *ceftriaxone*, which has an excellent response against all strains of the gonococcus bacterium,[12] is now the preferred treatment.[13] Ceftriaxone cures all forms of uncomplicated gonorrhea.[14] An alternative antibiotic such as *spectinomycin* may be used with people who cannot tolerate ceftriaxone. Successful treatment provides no protection against reinfection with gonorrhea if the person is again exposed to the bacterium. Sexual partners of persons with gonorrhea (or any other STD) should be evaluated by a physician, irrespective of whether any symptoms are present.

---

### HEALTH TIPS

▶ While some forms of gonorrhea are resistant to penicillin, other antibiotics are available that can cure all uncomplicated forms of gonorrhea.

---

Since gonorrhea and chlamydia (with or without symptoms) often occur together, persons infected with gonorrhea are usually also treated for chlamydia through the use of another antibiotic, generally *doxycycline*—or alternately *tetracycline* or *erythromycin*—administered orally over a course of seven days.[15]

---

### HEALTH TIPS

▶ Since the bacteria causing gonorrhea and chlamydia often travel together, if you have one, there's a good chance that you have the other. The good news is that antibiotic treatment effectively cures both.

---

## STD FACT SHEET: GONORRHEA

| WHAT IT IS | A bacterial STD |
|---|---|
| WHAT CAUSES IT | Gonococcus bacterium (*Neisseria gonorrhoeae*) |
| HOW IT IS TRANSMITTED | Vaginal or anal intercourse, oral-genital contact, or from mother to baby during childbirth |
| SIGNS AND SYMPTOMS | In men, a yellowish, thick penile discharge, burning urination<br><br>In women, increased vaginal discharge, burning urination, irregular menstrual bleeding (most women show no early symptoms) |
| HOW IT IS DIAGNOSED | Clinical inspection, culture of sample discharge |
| HOW IT IS TREATED | Antibiotics, typically ceftriaxone or spectinomycin |

## CHAPTER SEVEN

# WHAT YOU SHOULD KNOW ABOUT SYPHILIS

*DID YOU KNOW THAT?*

- ▶ *Increased rates of syphilis in our society are linked to the increased use of cocaine.*
- ▶ *There is no evidence that syphilis can be picked up from a toilet seat.*
- ▶ *You stand a one in three chance of contracting syphilis from a single sexual contact with an infected partner.*
- ▶ *Syphilis does not disappear when the initial symptoms abate.*
- ▶ *Untreated syphilis can eventually lead to brain and heart damage and result in death.*

No society wanted to be associated with **syphilis.** In Naples they called it "the French disease"; in France it was "the Neapolitan disease." Many Italians called it "the Spanish disease," but in Spain they called it "the disease of Española" (modern Haiti). But what is syphilis? What causes it? How is it spread? Why is syphilis on the rise in our society?

# What is Syphilis?

Syphilis is a sexually transmitted bacterial infection. *Treponema pallidum*, the bacterium that causes syphilis, was first isolated in 1905 by the German scientist Fritz Schaudinn. The name *Treponema pallidum* (*T. pallidum*, for short), reflects Greek and Latin roots meaning a "faintly colored (pallid) turning thread"—an apt description of the corkscrewlike organism when seen under the microscope (see Figure 7.1). Because of its spiral shape, *T. pallidum* is also called a *spirochete* from Greek roots meaning "spiral" and "hair."

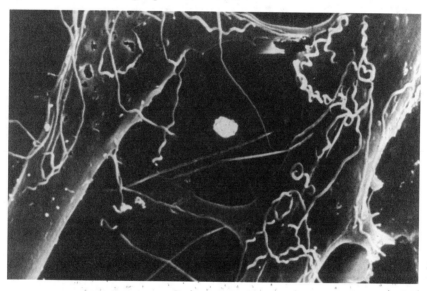

**Figure 7.1.** *Treponema Pallidum.* *Treponema pallidum* is the bacterium that causes syphilis. Because of the spiral shape, *T. pallidum* is also called a *spirochete*.

# How Widespread is Syphilis?

The incidence of syphilis decreased in the United States with the introduction of penicillin, an antibiotic which is used to treat the infection.[1] But despite the availability of penicillin, syphilis made a comeback in the U.S. during the 1980s. During the 1990s the rate of syphilis once again declined, reaching in 1995 the lowest level of reported cases since 1960.[2] Still, an estimated 120,000 new cases of syphilis occur annually in the U.S., though most cases are not reported to health officials.[3]

Researchers believe that the increase in syphilis during the 1980s was linked to the increased use of cocaine.[4] Cocaine users may be at greater risk of contracting syphilis than nonusers because they are more likely to engage in risky sexual practices, such as sex with multiple partners or with prostitutes, not because of their cocaine use per se.[5] Syphilis and other STDs are often spread among drug users, prostitutes (many of whom abuse drugs themselves), and their sex partners.[6] The problem is most acute in economically depressed inner-city areas, largely because such communities are disproportionately affected by problems of drug abuse and prostitution.

Though syphilis is less prevalent than gonorrhea, it can lead to more grievous effects, including blindness, heart disease, deteriorative mental illness, and even death.

# How is Syphilis Transmitted?

Syphilis is most often transmitted through vaginal or anal intercourse, or oral-genital or oral-anal contact with an infected person. Transmission usually occurs during contact between the open (infectious) lesions on the body of the infected partner and the mucous membranes or skin abrasions of the partner's body. The chances of contracting syphilis from one sexual contact with an infected partner is estimated at one in three.[7] While syphilis may be contracted by directly touching a **chancre,** there is no evidence that it can be transmitted by using the same toilet seat as an infected person.

```
╔══════════════════════════════════════════════════════╗
║                    HEALTH TIPS                         ║
╠══════════════════════════════════════════════════════╣
║  ►  You stand about a one in three chance of contracting syphilis from a
║     sexual contact with an infected person.
╚══════════════════════════════════════════════════════╝
```

Since the spirochete can cross the placental membrane, a pregnant woman can pass along the microbe to the fetus in-utero. Miscarriage, stillbirth, or **congenital syphilis** may result, causing impaired vision and hearing, or deformed bones and teeth. Diagnosis of syphilis in the mother by means of a blood test may help avoid transmission of the disease to the baby if treatment is administered early in pregnancy. The fetus will probably not be harmed if an infected mother is treated before the fourth month of pregnancy.

# What Are the Symptoms of Syphilis?

Syphilis progresses through several stages, each of which has characteristic symptoms. In the first or *primary stage*, a painless chancre (a hard, round, ulcerlike lesion with raised edges) appears at the site of infection two to four weeks after contact. When women are infected, the chancre usually forms on the vaginal walls or the cervix. It may also form on the external genitalia, most often on the labia (the folds of tissue that cover the vaginal opening). When men are infected, the chancre usually forms on the tip (glans) of the penis, or sometimes on the scrotum (external sac containing the testes or testicles) or penile shaft (the cylindrical body of the penis under the glans). If the mode of transmission is oral sex, the chancre may appear on the lips or tongue. If spread by anal sex, the rectum may serve as the site of the chancre. The chancre disappears within a few weeks but if the infection remains untreated, syphilis will continue to work within the body.

The *secondary stage* begins a few weeks to a few months later and is noted by the appearance of a skin rash consisting of painless, reddish raised bumps that darken after a while and burst, oozing a discharge. Other symptoms include sores in the mouth, painful swelling of joints, a sore throat, headaches, and fever, so a sufferer may wrongly assume that he or she has "the flu." A person may infect others during the primary stage, and for several years into the secondary stage. A pregnant woman, however, may pass along the infection to her newborn at any stage during the course of the infection.[8]

These symptoms soon disappear, and the infection enters the *latent stage* and may lie dormant for perhaps 1 to 40 years. But spirochetes continue to multiply during the latent stage and burrow into the circulatory system, central nervous system (brain and spinal cord), and bones.

In many cases, the disease eventually progresses to a late or *tertiary* stage. A large ulcer may form on the digestive organs, liver, lungs, skin, muscle tissue, or other organs. This destructive ulcer can often be successfully treated, but still more serious damage can occur as the infection attacks the cardiovascular system (heart and major blood vessels) or the central nervous system, in the latter case causing a condition called *neurosyphilis*. Neurosyphilis can cause brain damage, resulting in paralysis, or a deteriorating form of mental illness called **general paresis.** Damage to the cardiovascular system or central nervous system can also be fatal. The primary and secondary symptoms of syphilis inevitably disappear, so victims may be tempted to believe that they are no longer at risk. They may, therefore, fail to see a doctor. This is indeed unfortunate, because failure to eradicate the infection through proper treatment may eventually lead to the dire consequences associated with tertiary syphilis.

---

## HEALTH TIPS

▶ As with many other STDs, the disappearance of initial symptoms does not mean that the disease is cured. In the case of syphilis, the disease progresses through several stages if left untreated, and may eventually result in serious complications or even death.

# How is Syphilis Diagnosed?

Primary-stage syphilis is diagnosed by clinical examination. When a chancre is found, fluid may be drawn from it and examined under a microscope in a procedure called a dark-field test, which makes the spirochetes more visible to the eye. The most widely used blood test is the **VDRL.** The VDRL tests for the presence of antibodies in the blood to *Treponema pallidum.* Blood tests are usually not reliable until the secondary stage begins, as it may take some time after initial infection for antibodies to the bacterium to appear in the blood.

# How is Syphilis Treated?

Penicillin is the treatment of choice for syphilis. Primary and secondary syphilis, and latent syphilis of less than one year's duration, are generally treated with a single injection of penicillin (or doxycycline, tetracycline, or erythromycin for nonpregnant penicillin-allergic patients).[9] Latent stage syphilis of more than one year in duration, and neurosyphilis, are generally treated with multiple doses of injectable penicillin for two or three weeks. Like gonorrhea, successful treatment does not prevent reinfection. Also, like gonorrhea and other STDs, sex partners of persons infected with syphilis should also be evaluated by a physician.

## STD FACT SHEET:  SYPHILIS

| | |
|---|---|
| **WHAT IT IS** | A bacterial STD |
| **WHAT CAUSES IT** | The bacterium, *Treponema pallidum* ("spirochete") |
| **HOW IT IS TRANSMITTED** | Vaginal or anal intercourse, oral-genital contact, touching an infectious sore or chancre |
| **SIGNS AND SYMPTOMS** | Primary symptom is a hard, round painless sore (chancre) at the site of infection. If left untreated, the infection may progress through several stages. |
| **HOW IT IS DIAGNOSED** | In its primary stage, diagnosis is by clinical examination of the chancre and microscopic examination of fluid drawn from a chancre. A blood test, the VDRL, may be used if the disease reaches the secondary stage of infection. |
| **HOW IT IS TREATED** | Antibiotic treatment with penicillin, or doxycycline, tetracycline, or erythromycin for nonpregnant penicillin-allergic patients |

CHAPTER EIGHT

# WHAT YOU SHOULD KNOW ABOUT CHLAMYDIA

*DID YOU KNOW THAT?*

- *Chlamydia occurs more commonly than either gonorrhea or syphilis.*
- *Both men and women can develop chlamydial infections.*
- *NGU, an infection frequently traced to the chlamydia bacterium, occurs about two or three times as often in men as gonorrhea, although men are more likely to have heard of gonorrhea than NGU.*
- *If you have gonorrhea, there's a good chance you also have a chlamydial infection.*
- *The majority of men and women with chlamydial infections experience no symptoms and are generally unaware that they are infected.*
- *Untreated chlamydia can lead to the development of PID in women, which can result in infertility.*

What bacterial STD occurs most often—gonorrhea, syphilis, or chlamydia? Although you may be more familiar with gonorrhea and syphilis, chlamydial infections are more widespread.[1] Chlamydia is the fastest growing STD in the U.S., affecting as many as four million men and women annually.[2] Sexually active teenagers and college students constitute two of the highest risk groups for chlamydial infections.[3] Researchers estimate that between 8 and 40 percent of teenage women become infected.[4]

---

**HEALTH TIPS**

▶ People today stand a greater chance of contracting a chlamydial infection than either gonorrhea or syphilis. Yet fewer people are acquainted with chlamydia than with these other STDs.

---

# What is Chlamydia?

Chlamydia is a term used to describe various infections caused by the *Chlamydia trachomatis* bacterium, a unique parasitic organism that survives only within cells.[5] The particular type of infection depends upon the particular organs that are affected by the bacterium. Among the infections caused by *Chlamydia trachomatis* are *nongonococcal urethritis* (NGU) in men and women, *epididymitis* in men (infection of the epididymis), and *cervicitis* (infection of the cervix), *endometritis* (infection of the endometrium, the inner layer of the uterus), and PID (pelvic inflammatory disease) in women.[6]

# Urethral Infections and Chlamydia

Urethritis is a term used to refer to any type of inflammation or infection of the urethra, the canal through which urine is carried from the bladder to the outside world. Nongonococcal urethritis or NGU refers to any type of urethritis not caused by the gonococcal bacterium. (NGU was formerly called nonspecific urethritis or NSU.) While many organisms can cause NGU, *Chlamydia trachomatis* is the most common culprit,[7] accounting for about half of the cases among men.[8] NGU seems to be about two or three times as prevalent among men in the United States as gonorrhea.[9] Men in the 20 to 24 age range are considered most at risk for both gonorrhea and NGU,[10] presumably because of the high level of sexual activity among men in this age range. Teenage men 15 to 19 years of age and men in their late twenties are the two next highest risk groups.[11]

NGU is generally diagnosed only in men. Women with urethritis due to *Chlamydia trachomatis* are generally said to have chlamydia or a chlamydial infection.

Chlamydial infections also frequently occur together with other STDs, most often gonorrhea. As many as 45 percent of cases of gonorrhea involve coexisting chlamydial infections.[12]

# What Are the Symptoms of Chlamydia?

Chlamydial infections may produce symptoms that are similar to those of gonorrhea, but are usually milder. NGU in men may produce a thin, whitish discharge from the penis and some burning or other pain during urination. These symptoms contrast with the yellow-greenish discharge and more intense pain associated with gonorrhea. There may be feelings of soreness and heaviness in the testes. A chlamydial infection can progress to infect the epididymis, causing epididymitis, which is associated in young men with such symptoms as pain, swelling and tenderness of the scrotum, and fever.[13]

---

**HEALTH TIPS**

▶ The symptoms of chlamydia are similar to those of gonorrhea, but are generally milder.

---

Women with a chlamydial infection may experience burning during urination, genital irritation, and abnormal vaginal discharge. Women may also experience pelvic pain and irregular menstrual cycles. The cervix may look swollen and inflamed. Oral sex performed on an infected partner may lead to the development of a throat infection in either gender.

# Chlamydia Often Occurs without Symptoms

About three of four cases of chlamydial infection in women and about one in four in men have no noticeable symptoms.[14] For this reason, chlamydia is often referred to as a "silent disease." People with asymptomatic chlamydial infections can innocently pass along their infections to other sex partners.

---

**HEALTH TIPS**

▶ Most people with chlamydial infections, perhaps two in three, have no symptoms. Yet chlamydia can lead to serious consequences, and be passed along to others, whether or not it produces any noticeable symptoms.

---

# How is Chlamydia Transmitted?

Chlamydial infections are usually transmitted sexually through vaginal or anal intercourse. If a person touches his or her eyes after handling the genitals of an infected partner, chlamydia can also cause an eye infection. Chlamydia can also infect the throat if one has oral-genital sex with an infected partner.[15]

Chlamydial infections frequently have nonsexual origins in underdeveloped countries,[16] where they are often spread by contact with fecal matter or with discharges on the fingers or skin of infected persons, or by insect bites. In industrialized countries like the U.S. and Canada, however, virtually all cases of Chlamydial infections in adults are transmitted sexually.[17]

Newborns can acquire potentially serious chlamydial eye infections as they pass through the birth canal of infected mothers during birth. Even newborns delivered by Caesarean section (also called C-section) may be infected if the amniotic sac (the sac holding the fetus) breaks before delivery.[18] Each year, more than 100,000 infants are infected with the bacterium during birth.[19] Of these, about 75,000 develop eye infections (conjunctivitis) and 30,000 develop a form of pneumonia caused by the chlamydial bacterium. Evidence from several studies suggests that between 2 and 26 percent of pregnant women in the United States carry *Chlamydia trachomatis* in their cervix, and are thus capable of transmitting the organism to their newborns upon delivery.[20]

# What Are the Risks Associated with Untreated Chlamydia?

Because chlamydia is often asymptomatic, many infected people go untreated. In untreated women, a chlamydial infection, like a gonorrheal infection, can spread to involve the reproductive tract, possibly leading to PID and scarring of the fallopian tubes, which can result in infertility.[21] Researchers estimate that perhaps half of the more than one million cases of PID diagnosed each year are attributable to chlamydial infections.[22] (Untreated gonorrhea is also a principal culprit.) PID can also lead to an ectopic (tubal) pregnancy. Women with a history of exposure to *Chlamydia trachomatis* stand twice the chance of developing an ectopic pregnancy.[23]

Untreated chlamydial infections can also spread internally in men, causing such conditions as epididymitis. Chlamydial infections account for about half of the cases of epididymitis.[24] Swelling, feelings of tenderness, and pain in the scrotum are the principal symptoms. Fever may also be present. Yet only about one or two percent of men with untreated NGU caused by *Chlamydia trachomatis* go on to develop epididymitis.[25] That's no excuse for not seeking treatment, but men do stand a much lesser chance of serious complications from chlamydial infections than do women. Still, evidence concerning the long-term effects of untreated chlamydial infections in men remains inconclusive.

---

### HEALTH TIPS

▶ In women, untreated chlamydial infections can lead to PID and result in sterility, as well as increase the risk of tubal pregnancies; in men, the long-term consequences of untreated chlamydial infections remain undetermined.

---

# How Are Chlamydial Infections Diagnosed?

A laboratory test, the Abbott Testpack, permits physicians to verify a diagnosis in women in about 25 or 30 minutes.[26] The test analyzes a cervical smear (like a Pap smear) and identifies about 75 to 80 percent of infected cases, with relatively few *false positives* (incorrect positive findings). In men, a swab is inserted through the penile opening and the extracted fluid is analyzed to detect the presence of *Chlamydia trachomatis*.

# How Are Chlamydial Infections Treated?

Antibiotic treatment with a seven-day course of doxycycline or a single dose of azithromycin is the recommended treatment.[27] Treatment is highly effective in curing the infection. An alternative antibiotic, erythromycin, is recommended for pregnant women. (Penicillin, effective in treating gonorrhea, is ineffective against *Chlamydia trachomatis*.) A person does not acquire immunity to chlamydia as the result of having had the infection or being successfully treated for an infection; hence reinfection is possible following a subsequent exposure to the bacterium.

Evaluation and treatment of sex partners is considered critical regardless of apparent symptomatology to prevent the infection from being passed back and forth.[28] Sex should be avoided until patients and their partners complete treatments and show an absence of symptoms. A woman whose sex partner develops NGU should be examined for a chlamydial infection; at least 30% of these women will test positive for a chlamydial infection even though they (and their partners) may be free of symptoms.[29]

Men whose sex partners develop urethral or cervical infections should be medically evaluated, whether or not they notice any symptoms themselves. With chlamydia, both partners may be unaware that they are infected, and may be oblivious to the damage the infection is causing internally. Because of the risks posed by untreated chlamydial infections, especially to women, and the high rate of asymptomatic infections, routine screening of sexually active young women during regular gynecological examinations is recommended irrespective of presenting symptoms.[30]

---

**HEALTH TIPS**

▸ If your sexual partner has a chlamydial infection, you should be tested too, whether or not you have any symptoms.

---

## STD FACT SHEET: CHLAMYDIA

| WHAT IT IS | A bacterial STD |
| --- | --- |
| WHAT CAUSES IT | The bacterium, *Chlamydia trachomatis* |
| HOW IT IS TRANSMITTED | Vaginal or anal intercourse or oral-genital contact; touching the eyes after touching the genitals of an infected partner; from infected mother to newborn during childbirth |
| SIGNS AND SYMPTOMS | Women may notice more frequent urination, which may be painful. Lower abdominal pain and pelvic inflammation may occur, as well as a vaginal discharge. Most women, however, are asymptomatic.<br><br>Men may experience symptoms similar to gonorrhea, but generally milder, such as burning or painful urination and a slight penile discharge. Men may also be symptom-free. Either gender may experience a sore throat following oral-genital contact with an infected partner. |
| HOW IT IS DIAGNOSED | In women, analysis of a cervical smear; in men, analysis of an extract of fluid drawn from the penis. |
| HOW IT IS TREATED | Use of antibiotics, such as doxycycline, tetracycline, or erythromycin. |

## CHAPTER NINE

# WHAT YOU SHOULD KNOW ABOUT VAGINAL INFECTIONS

## *DID YOU KNOW THAT?*

▸ *Men, too, can be infected by the organisms causing vaginal infections in women.*

▸ *Vaginitis is often caused by an overgrowth of infectious organisms that normally reside in the vagina.*

▸ *Vaginal yeast infections are actually caused by a fungus.*

▸ *According to a recent study, recurrent vaginal yeast infections can be reduced by eating a pint a day of yogurt that contains active bacterial cultures.*

▸ *The fungus that causes vaginal yeast infections may be transmitted by contact with a moist towel used by an infected woman.*

▸ *Three of four men whose partners have "trich" are found to be infected themselves, many unknowingly.*

▸ *Both partners with "trich" should be treated simultaneously, even if they have no symptoms, so as to avoid bouncing the infection back and forth between them.*

Men may think that this chapter doesn't concern them and be tempted to start turning the pages at this point. While it is true that only women can suffer from vaginal infections, the microbes that cause these annoying infections may infect the male's urethral tract and be passed back and forth between sexual partners. So men, please, stay tuned. The information contained here concerns you as well.

---

### HEALTH TIPS

▶ While only women can have vaginal infections (naturally, since only women have vaginas), the microbes causing these infections may also cause problems for men.

---

# What is Vaginitis?

**Vaginitis** is a general term that refers to any type of vaginal inflammation or infection. Different types of vaginitis are caused by different microorganisms. Most cases of vaginitis,[1] perhaps as many as 90 percent, involve bacterial vaginosis (formerly called *nonspecific vaginitis*), candidiasis (commonly called a "yeast" infection), or trichomoniasis ("trich"). Some cases involve combinations of the three. Bacterial vaginosis is the most common form of vaginitis, followed by candidiasis, then by trichomoniasis.[3]

# What Causes Vaginitis?

Some cases of vaginitis are caused by infectious organisms that are spread by sexual contact. But other cases of vaginitis result from an allergic reaction or sensitivity to certain chemicals, rather than sexual transmission. In still other cases, organisms that ordinarily reside in the vagina without causing problems become infectious when changes in the vaginal environment allow them to multiply and overgrow.

## Factors that Increase the Risk of Vaginal Infections

The environmental balance of the vagina may be upset and increase the chances of a vaginal infection due to such factors as excessive douching, dietary changes, use of antibiotics or birth control pills, lowered resistance (perhaps from fatigue or poor diet), changes in the natural body chemistry, or even wearing pantyhose or nylon underwear.

---

### HEALTH TIPS

▸ Various factors may be involved in changing the natural flora of the vagina, leading to an overgrowth of infectious organisms that may result in vaginitis.

---

# What is Bacterial Vaginosis (BV)?

Bacterial vaginosis (BV) is caused by various bacteria, most commonly the bacterium *Gardnerella vaginalis*.[4] The infection arises when the bacteria causing bacterial vaginosis grow in place of the bacterial flora that normally reside in the vagina.[5]

## What Are the Symptoms of BV?

The most prominent symptom in women is a thin, foul-smelling vaginal discharge. Yet many cases occur without symptoms. Besides causing troublesome symptoms in some cases, medical authorities have raised the possibility that BV may increase the risk of various gynecological problems, including infections of the reproductive tract.[6]

---

### HEALTH TIPS

▸ Bacterial vaginosis is not merely annoying, but may be linked to more serious problems. Women should report any noticeable symptoms to their doctors.

---

## How is BV Transmitted?

The bacterium is often transmitted sexually between partners, usually during sexual intercourse. Men may harbor *Gardnerella vaginalis* in their urethras (the tube from the bladder through the penis through which urine passes) which they presumably contracted from sexual contact with infected female partners.[7]

## How is BV Diagnosed?

An accurate diagnosis depends on culturing and identifying the bacteria in the laboratory.[8]

# What is Candidiasis?

Candidiasis or *monilias*, which is more commonly referred to as a vaginal yeast infection, is a vaginal infection caused by *Candida albicans*, a yeast-like fungus. The fungus is normally found in small quantities in the vagina. It usually causes no symptoms when the vaginal environment is normally balanced, but becomes infectious when the normal balance of the vaginal environment is upset, allowing an overgrowth of the fungus to occur. Yeast infections can also occur in the mouth in both men and women and in the penis in men. Estimates suggest that about three in four women will experience a vaginal yeast infection during their lives and perhaps 40 to 50 percent of women who do will encounter at least one recurrence.[9]

Frequent or persistent vaginal yeast infections are sometimes an early warning sign of HIV infection.[10] Stubborn or repeated cases of vaginal yeast infections are the most common initial symptoms of HIV infection in women. *However, we should underscore the fact that the great majority of women with recurrent vaginal yeast infections are not infected with HIV.*

**HEALTH TIPS**

▶ Vaginal yeast infections are exceedingly common, affecting perhaps three in four women at one time or another. They also frequently recur, although women may take certain precautions that can reduce the risk of recurrent infections.

## What Are the Symptoms of Vaginal Yeast Infections?

Candidiasis commonly produces soreness, inflammation, and intense (sometimes maddening!) itching in the genitals that is accompanied by a white, thick, curdy vaginal discharge.

## How is a Yeast Infection Diagnosed?

Diagnosis is usually made on the basis of symptoms. A microscopic examination of vaginal fluids may be used to detect the presence of yeast cells. Culturing a sample of vaginal secretions is a yet more sensitive means of detecting yeast cells.[11]

## What Factors Lead to an Overgrowth of "Yeast"?

The use of antibiotics or birth control pills, pregnancy, and diabetes are frequently implicated as factors accounting for changes in the vaginal balance that allow the fungus that causes yeast infections to proliferate. So, too, is wearing tight-fitting, restrictive, and poorly ventilated clothing or nylon underwear.

---

**HEALTH TIPS**

---

Factors that may increase the risk of vaginal yeast infections include:
- ▶ Use of antibiotics
- ▶ Use of birth control pills
- ▶ Pregnancy
- ▶ Diabetes
- ▶ Wearing constricting, tight-fitting clothing or nylon underwear

---

**Does Diet Play a Role?**   There is increasing evidence that diet may play a role in recurrent yeast infections.[12]   Researchers have recently reported that daily ingestion of one pint of yogurt containing active bacterial (*Lactobacillus acidophilus*) cultures may reduce the rate of recurrent infections.[13]

---

**HEALTH TIPS**

---

- ▶ Research evidence shows that eating one pint a day of yogurt with active bacterial cultures may reduce the risk of recurrent vaginal yeast infections.

---

## Can the Fungus That Causes Yeast Infections Be Passed Back and Forth Between Sexual Partners?

Yes, the fungus that causes yeast infections may be transmitted sexually during sexual intercourse, or even passed back and forth between sex partners. However, the role of sexual contact in spreading the infection is believed to be limited.[14] (Most infections in women are believed to be caused by an overgrowth of "yeast" normally found in the vagina, not by sexual transmission).

## Can a Yeast Infection Be Transmitted Nonsexually?

Yes, the fungus that causes yeast infections may also be passed nonsexually, such as by sharing a moist towel or washcloth with an infected woman.[15]

---

## HEALTH TIPS

▶ Never share a towel with a person who has a vaginal yeast infection. For that matter, since you can never know who is carrying an STD, it's prudent to avoid sharing towels or other personal articles with anyone. Period.

---

## Can Men Become Infected by the Organism That Causes Vaginal Yeast Infections in Women?

Yes, *Candida albicans*, the organism responsible for vaginal yeast infections, can cause urinary infections in men. *Candida* can be found in fluid extracted from the penises of about one in five male sexual partners of women with recurrent yeast infections. While most men with *candida* are asymptomatic,[16] some may develop NGU or a genital thrush which is accompanied by sensations of itching and burning during urination, or a reddening of the penis.[17]

---

## HEALTH TIPS

▶ Men may become infected with *Candida albicans*, the organism causing vaginal yeast infections in women. In men, the infection often takes the form of NGU, genital thrush, or a reddening of the penis.

---

## Can Candidiasis Be Spread to the Mouth or Anus?

Yes, oral and anal sex can be the means of transmission of candidiasis from the genitals of one partner to the mouth and anus of the other, respectively, in either gender.

---

## HEALTH TIPS

▸ Candidiasis may appear in other parts of the body than the genitals, such as the mouth and anus, as the result of oral-genital and anal-genital contact, respectively.

---

# What is Trichomoniasis?

**Trichomoniasis** or "trich" (pronounced *trick*) is caused by a protozoan (one-celled organism) type of parasite called *Trichomonas vaginalis*. Trichomoniasis is the most common parasitic STD,[18] accounting for some eight million cases annually among women in the United States.[19] "Trich" in the female is characterized by burning or itching in the genitals and the appearance of a foamy whitish to yellowish-green discharge that is often odorous. Lower abdominal pain is reported by 5 to 12 percent of infected women.[20] Upwards of half of infected woman experience a mild degree of pain during sexual intercourse or upon urination.[21] Many women notice symptoms appearing or worsening during, or just following, their menstrual periods.[22] Trichomoniasis has been linked to the development of tubal adhesions that can result in infertility.[23]

## How is "Trich" Diagnosed?

Examination under a microscope of a smear of vaginal secretions may be used to confirm a diagnosis in the doctor's office.[24] Yet a more sensitive method of confirming a diagnosis is based on examination of cultures grown from a sample of the woman's vaginal secretions.[25]

## Can Men Get "Trich"?

In the male, *Trichomonas vaginalis* can lead to NGU, which may be noticeable by the appearance of a slight penile discharge (usually only upon first urination following morning awakening). The urethra may become slightly irritated, leading to sensations of itching or tingling along the urethral tract. However, most infected men are asymptomatic.[26] Evidence shows that three or four of ten male partners of women with "trich" harbor the infectious organism in their urinary tract,[27] in many cases unknowingly.

## Can "Trich" Occur without Symptoms in Women Too?

Yes, "trich" occurs without symptoms in about one woman in two.[28] Regardless of symptoms, the infection can be passed along to one's sexual partners. "Trich" often occurs together with other STDs, such as gonorrhea, so it is important for a person with "trich" to be checked out for other STDs as well.[29]

---

### HEALTH TIPS

▸ Most of the men, and about half of the women, infected with "trich" experience no symptoms. Frequently, men and women pass the infection back and forth without realizing it.

---

## Can "Trich" Be Transmitted Nonsexually?

In relatively few cases, "trich" may be transmitted by nonsexual means. Since the organism can survive for a few hours in bodily fluids deposited outside the body, it may be passed along in semen, vaginal secretions, or urine found on damp towels washcloths, or bedclothes. There is a slight possibility that "trich" may be contracted from contact with a toilet seat used by an infected person, but only if the organism makes direct contact with the woman's genitals or the man's penis.[30]

# What You Should Know about Treating Vaginal Infections

## How is Bacterial Vaginosis (BV) Treated?

The drug metronidazole (brand name Flagyl) is effective in treating bacterial vaginosis; however, the drug should not be used with women during the first trimester of pregnancy.[31] The medication is administered orally, typically over the course of seven days.[32] It remains unclear whether a man whose partner is infected with bacterial vaginosis should also be treated. Most men harbor the bacterium in their urethral tract without any symptoms.[33] Lacking symptoms, they may unknowingly transmit the bacterium to their sexual partners. Yet firm evidence is lacking to show that treatment of the male partner reduces the risk of recurrence in the female.[34] Nonetheless, women who suffer from bacterial vaginosis should consult with their physicians about the advisability of treating their sexual partners.

## How Are Vaginal Yeast Infections Treated?

The drugs miconazole (brand name Monistat), clotrimazole (brand names Lotrimin and Mycelex), and terconazole (brand name Terazol) are effective in treating vaginal yeast infections.[35] Some of these medications are now available without a prescription. Even so, it would be wise for women who encounter any vaginal complaints to consult a physician before starting any medication to ensure that they receive the appropriate diagnosis and treatment.

Vaginal infections are frequently recurrent. Successful treatment of vaginal infections does not bestow immunity and recurrences occur commonly. Nor is there evidence showing that treatment of the male partner prevents recurrence in the female partner.[36]

## How is "Trich" Treated?

Metronidazole (Flagyl) is also effective in treating trichomoniasis. Because "trich" may pass back and forth between sexual partners, both partners should be treated simultaneously, whether or not they report symptoms. Treatment is 90 to 100 percent effective when both sex partners are treated simultaneously.[37]

```
┌─────────────────────────────────────────────────────────────┐
│                        HEALTH TIPS                            │
├─────────────────────────────────────────────────────────────┤
│  ▸  While some medications for treating vaginal yeast         │
│     infections are available without a prescription, women    │
│     who suspect they are infected should consult a            │
│     physician to ensure that they receive the proper diag-    │
│     nosis and treatment.                                      │
└─────────────────────────────────────────────────────────────┘
```

# How to Reduce the Risk of Vaginitis

The Boston Women's Health Book Collective offers the following advice to women to help them reduce the risks of developing vaginitis:[38]

1. Wash your external genitalia and anus regularly with mild soap. Pat dry, being careful not to touch the genitals after dabbing the anus.
2. Wear cotton panties instead of nylon underwear since nylon retains heat and moisture that cause harmful bacteria to flourish.
3. Avoid pants that are tight in the crotch.
4. Be certain that sexual partners are well-washed. Use of condoms may also reduce the spread of infection from one's sexual partner.
5. Use a sterile, water-soluble jelly like K-Y jelly if artificial lubrication is needed for intercourse—not *Vaseline*. Birth control jellies can also be used for lubrication.
7. Avoid intercourse that is painful or abrasive to the vagina.
8. Avoid diets high in sugar and refined carbohydrates since they may alter the normal acidity of the vagina.
9. Women who are prone to vaginal infections may find it helpful to occasionally douche with plain water, a solution of 1 or 2 tablespoons of vinegar in a quart of warm water, or a solution of baking soda and water. Douches consisting of unpasteurized plain (unflavored) yogurt may help replenish the "good" bacteria that is normally found in the vagina and that may be destroyed by use of antibiotics. Be careful when douching, and do not douche when pregnant or when you suspect you may be pregnant. To be safe, you should consult your physician before douching or applying any preparations to the vagina.
10. Remember to take care of your general health. Eating poorly or getting insufficient rest will reduce your resistance to infection.

## STD FACT SHEET: VAGINITIS

| WHAT IT IS | A general term that applies to any type of vaginal infection. The major types are bacterial vaginosis, candidiasis (yeast infection), and trichomoniasis ("trich"). |
|---|---|
| WHAT CAUSES IT | Bacterial vaginosis: *Gardnerella vaginalis* and other bacteria. Candidiasis: the yeast-like fungus *Candida albicans*. Trichomoniasis: the parasite *Trichomonas vaginalis*. |
| HOW IT IS TRANSMITTED | May result from an allergic reaction, a change in the vaginal flora that allows an overgrowth of infectious organisms, or by sexual contact. Yeast infections may also be passed from one woman to another through the sharing of damp towels or washcloths. |
| SIGNS AND SYMPTOMS | Various signs, consisting mostly of irritation and itching of the genitals and by a foul-smelling vaginal discharge. Men infected with these agents may notice some itching and burning sensations during urination or perhaps a reddening or inflammation of the penis. Bacterial vaginosis and trichomoniasis are often asymptomatic in both men and women. |
| HOW IT IS DIAGNOSED | Clinical inspection of symptoms, which may be followed by microscopic examination of a cultured vaginal smear. |
| HOW IT IS TREATED | Oral administration of the drug metronidazole (Flagyl) for bacterial vaginosis and trichomoniasis; vaginal creams, suppositories or tablets containing the drugs miconazole, clotrimazole, or terconazole for candidiasis. |

## CHAPTER TEN

# WHAT YOU SHOULD KNOW ABOUT HERPES

## *DID YOU KNOW THAT?*

- *There are different forms of herpes caused by distinct, but related viruses.*
- *Some herpes infections of the genitals are actually caused by the oral herpes virus; some herpes infections of the mouth are caused by the genital herpes virus.*
- *Genital herpes may be contracted through sexual contact with an infected partner even during symptom-free periods.*
- *Although the symptoms of genital herpes disappear after an active episode, the virus remains in the body indefinitely and can cause recurrent outbreaks.*
- *Herpes outbreaks may be preceded by certain warning signs.*
- *A person's attitudes play an important role in determining the emotional consequences of coping with herpes.*

The hysteria that surrounded the rapid spread of herpes in the 1970s and early 80s has all but died since the world learned to dread an even more threatening STD: AIDS. Before AIDS, it was genital herpes that many sexually active people feared the most. Herpes was feared so much because it is incurable and, at the time, essentially untreatable. Herpes still remains a very real threat. Once you get herpes, it's yours for life. After the initial attack, it remains an unwelcome guest in your body, finding a cozy place to lie low until it stirs up trouble again, causing recurrent outbreaks that usually happen at the worst times, like around final exams. Not only can you not get rid of the virus, but you can pass it along to whomever else you have sex with for the rest of your life. Ugh! Relationships, and even many prospective marriages, have been zapped when one partner learns that the other is infected. But many people learn to cope with herpes and go on with their lives and loves, adapting to it as more of an occasional nuisance or annoyance than a wholesale catastrophe.

While much of the public alarm about herpes has been overshadowed by concerns about AIDS, herpes remains a troubling and potentially even a dangerous disease, as it appears to increase the risk of cervical cancer in women. While we cannot cure it, or rid the body of the virus that causes it, progress has been made in treating the disease. Here's what you should know about herpes:

# What is Herpes?

The term herpes refers to several different types of infections caused by related forms of the *herpes simplex* virus. The most common herpes virus, **herpes simplex virus type 1,** or HSV-1 virus, causes *oral herpes,* which appears as cold sores or fever blisters on the lips or mouth. Many people have recurrent "cold sores" without realizing that these are caused by a herpes virus. The oral herpes virus can also be transferred to the genitals by the hands or by oral-genital contact. It is believed that between 10 and 50 percent of the cases of genital herpes are actually caused by the HSV-1 virus (a kind of "cold sore" of the genitals).[1]

**Genital herpes** is caused by a related but distinct virus, the **herpes simplex virus type 2** (HSV-2). This virus produces painful shallow sores and blisters on the genitals. HSV-2 can also be transferred to the mouth through oral-genital contact. Thus either HSV-1 or HSV-2 may cause herpes outbreaks on the mouth and lips, or on the genitals. Since both viruses can be transmitted sexually, the diseases they cause can be classified as STDs. Both types of herpes viruses may also cause an infection of the throat. Oral sex is a frequent channel for transmitting the herpes virus (either HSV-1 or HSV-2) from the genitals to the throat.

# How Many People Have Herpes?

Estimates are that about 40 million people in the U.S. are infected with genital herpes.[2] About 500,000 new cases are believed to occur annually. Approximately two of three people with herpes don't know they are infected, either because they have no symptoms or the symptoms they do have are so mild they go unnoticed.[3] Even so, they are capable of infecting their sexual partners. More than 100 million Americans are believed to be infected with oral herpes.

## Who is Most at Risk?

With perhaps 40 million people in the U.S. infected with genital herpes, and perhaps more than 100 million infected with oral herpes, the chances of contracting one or both forms of herpes from potential sexual partners is fairly high, especially if one has sexual contact with multiple partners. The chance of contracting genital herpes from a single sexual encounter with an infected partner during a flare-up of the disease is estimated at about 50 percent for men and 80 to 90 percent for women.[4]

# How is Herpes Transmitted?

The herpes viruses can be transmitted through oral, anal, or vaginal sex with an infected person. Oral herpes is also easily contracted by drinking from the same cup, by kissing, and even by sharing moist towels with an infected person. Genital herpes is most often spread by vaginal or anal intercourse or by oral sex. Herpes viruses can also survive for several hours on toilet seats or other objects, where they may be transmitted by direct contact with the genitals, although this mode of transmission is probably unusual. Use of a latex condom may reduce the risk of transmitting or contracting genital herpes, especially if combined with a spermicide (a sperm-killing chemical, used for contraception) containing the ingredient nonoxynol-9, which not only deactivates sperm but is toxic to the herpes virus as well as to many other STD-causing organisms. Yet transmission may occur when herpes sores (lesions) are present even with the use of condoms.[5] It may be wise to err on the side of caution and abstain from intimate genital contact during an active episode of genital herpes, at least until any herpes sores completely heal over. People may also acquire the virus from sexual contact with an infected partner but never experience outbreaks themselves, yet still be capable of passing along the virus to others.[6]

+---

### HEALTH TIPS

| ▶ Since the HSV-2 virus cannot pass through a latex condom,[7] couples who are sexually active during herpes outbreaks should use a latex condom along with a spermicide that contains the ingredient nonoxynol-9, which is toxic to the herpes virus. | ▶ However, since use of a condom is no guarantee against transmission of herpes viruses (condoms may tear or fall off, or not cover all exposed areas that may come into contact with active lesions), it would be wise to avoid genital contact until herpes sores have completely healed. |

## Can Genital Herpes Be Transmitted Between Flare-ups?

Unfortunately, yes. While genital herpes is most contagious during active flare-ups of the disease, alarming new evidence shows that herpes may also be transmitted from an infected partner when no symptoms (genital sores or feelings of burning or itching in the genitals) are present.[8] Some researchers suspect that *all* infected women may shed some of the virus at least some of the time they are symptom-free, making it possible for them to unknowingly infect their partners at these times.[9] The relative risks of contracting genital herpes from a symptom-free infected partner remain unclear.

## Can Women Pass Along Herpes to Their Newborns?

Sadly, yes. An infected women may pass along genital herpes to her newborn as the baby makes its way through the birth canal during childbirth. Genital herpes is a potentially serious problem for a newborn, causing damage or even death.[10] To avoid the risk of the newborn becoming infected when passing through the birth canal, obstetricians usually perform Caesarean sections (C-sections) if the mother shows an active outbreak at the time of delivery. C-sections are currently recommended for women who show active lesions or warning signs of an infection at the time of delivery.[11] Unfortunately, the absence of symptoms may not guarantee the safety of the baby. In one study of some 16,000 pregnant women, a small proportion (about 3.5 women in a thousand) showed no symptoms of genital herpes but produced positive test results that indicated an active infection. They may thus have been capable of unknowingly infecting their newborns.[12]

### Can Herpes Be Spread From One Part of the Body to Another?

Herpes may be spread from one part of the body to another by touching the infected area and then touching or rubbing another body part. One potentially serious result is a herpes infection of the eye—**ocular herpes.** People with active oral herpes who wear contact lenses need to be careful not to put their lenses in their mouths to wet them before inserting them into their eyes, lest they risk spreading the infection to their eyes.

---

### HEALTH TIPS

▸  Although thorough washing with soap and water after touching an
   infected area may reduce the risk of spreading the infection to other
   parts of the body, it is better to avoid touching the infected area
   altogether, especially if active sores are present.

---

# Health Risks Associated with Genital Herpes

While herpes sores heal naturally, researchers find that herpes may lead to serious complications, especially in women. The miscarriage rate among herpes sufferers is more than three times higher than the rate for the general population.[13] Herpes also appears to place women at greater risk of genital cancers, such as cervical cancer, a potential killer.[14] All women, not just herpes sufferers, are advised to have regular pelvic examinations, including Pap smears for early detection of cervical cancer.

# What Are the Symptoms of Genital Herpes?

Genital herpes is characterized by the appearance of genital lesions or sores, which appear about six to eight days after infection. At first they appear as reddish, painful bumps, or papules (pimples) along the penis in the man or the external genitalia in the woman. The papules may also appear on the thigh or buttocks in men or women, or in the vagina or on the cervix in women. These papules turn into groups of small blisters that are filled with fluid containing infectious viral particles. The blisters are attacked by the body's immune system (white blood cells). They become filled with pus and break open, becoming extremely painful, shallow sores, or ulcers, that are surrounded by a red ring. The person is especially infectious at this time, as the ulcers shed millions of viral particles. Other symptoms may include headaches and muscle aches, swollen lymph glands, fever, burning on urination, and a vaginal discharge in women. The blisters crust over and heal in from one to three weeks. Internal sores in the vagina or on the cervix may take ten days longer than external (labial) sores to completely heal, so physicians advise infected women to avoid unprotected intercourse for at least 10 days following the healing of external sores.

| HEALTH TIPS | |
|---|---|
| ▸ Treat any appearance of a sore or blister around the genitals as a warning sign. Check it out with a physician. | ▸ Women with herpes should avoid unprotected intercourse for at least ten days following the healing of external sores, just in case there are any internal sores that take longer to heal. |

## Is Genital Herpes Cured if the Sores Heal on Their Own?

Although the symptoms of genital herpes disappear on their own within a few days or weeks, the disease does not. The virus remains in the body permanently, burrowing into nerve cells in the base of the spine, where it may lie dormant for years, or even a lifetime. The infected person is least contagious during this dormant stage. For reasons that remain unclear, the virus becomes reactivated and gives rise to symptoms in 30 to 70 percent of cases.[15] On the other hand, some people, perhaps about 10 percent,[16] have no recurrences; for many others, recurrences are milder and briefer in duration than initial episodes and become less frequent over time.

## What Factors Might Prompt Recurrences?

Though the causal factors explaining recurrences are not entirely clear, scientists suspect that recurrences may be related to such factors as infections (as in a cold), stress, depression, exposure to the sun, and hormonal changes, such as those that occur during pregnancy or menstruation.[17] Yet not all researchers have been able to link emotional stress to herpes flare-ups.[18] Recurrences tend to occur within three to twelve months of the initial episode and to affect the same part of the body. They tend to be milder, and briefer in duration, than initial episodes, lasting from about three days to two weeks, and are often asymptomatic. Both initial and recurrent episodes of genital herpes tend to produce more severe symptoms in women, such as painful genital lesions.[19] Recurrences often decrease in frequency over time and may eventually disappear. Flare-ups sometimes continue to recur, however, with annoying frequency over the course of a lifetime.

---

### HEALTH TIPS

**To Reduce the Risk of Recurrent Herpes:**

- Maintain regular sleeping habits.
- Avoid unnecessary stress.
- Learn to manage the stress you can't avoid.
- Avoid excessive exposure to the sun.

- Take good care of your general health.
- Take good care of your mental health and well-being.

---

## Are there Warning Signs of an Impending Outbreak?

About 50 percent of people with recurrent genital herpes experience warning signs (called prodromal symptoms) before an active outbreak.[20]   These may include feelings of burning, itching, pain, tingling, or tenderness in the affected area.   These symptoms may be accompanied by sharp pains in the lower extremities, groin, or buttocks.   Herpes sufferers may be more infectious when prodromal symptoms appear and should avoid unprotected sex until the flare-up is resolved.   Infectiousness escalates with the appearance of active sores.   Sometimes the symptoms are so mild as to go unnoticed, so that people are unaware of being infectious.

---

### HEALTH TIPS

▸   To reduce the risk of infecting their sexual partners, people with herpes should avoid unprotected sex beginning with the first appearance of any warning signs of an impending recurrence and lasting through the course of an active episode.

---

# What Are the Symptoms of Oral Herpes?

The symptoms of oral herpes include sores or "fever blisters" on the lips, the inside of the mouth, the tongue, or the throat.   Fever and feelings of sickness may occur. The gums may become red and swollen.   The sores heal over in about two weeks, and the virus retreats into nerve cells at the base of the neck, where it lies dormant between flare-ups.   About 90 percent of oral herpes sufferers experience recurrences, and about half of these have five or more recurrences during the first two years after the initial outbreak.[21]

# How is Herpes Diagnosed?

Herpes is first diagnosed by clinical inspection of the herpes sores or ulcers in the mouth or on the genitals.   In the case of genital herpes, a sample of fluid may be taken from the base of a genital sore and cultured in the laboratory to detect the growth of the virus.[22]

# How is Herpes Treated?

There is no cure nor safe and effective vaccine for herpes. Viruses, unlike the bacteria that cause gonorrhea or syphilis, do not respond to antibiotics. Antiviral drugs such as *acyclovir* (brand name Zovirax) and valacyclovir (brand name Valtrex) can help relieve pain, speed healing, and reduce the duration of viral shedding (the period of time during which the virus is found in vaginal secretions and semen) in genital herpes when applied directly in ointment form to the herpes sores. Acyclovir must be administered orally, in pill form, to be effective against internal lesions in the vagina or on the cervix. Oral administration of acyclovir may reduce the severity of the initial episode and, if taken regularly, the frequency and duration of recurrent attacks of genital herpes.[23] In a recent study of 389 people who had suffered 12 or more recurrences annually, researchers found that daily doses of acyclovir reduced the number of recurrences, on the average, to 1.7 in the first year, and to less than 1 (0.8) by the fifth year, with few adverse reactions reported.[24] In another study of 525 genital herpes sufferers over a three-year period, regular use of oral acyclovir reduced the frequency of outbreaks from an average of 12 per year to only 1 or 2 per year.[25] Yet the safety of long-term use of acyclovir needs further study. Nor has the safety of using acyclovir during pregnancy been established.[26]

---

### HEALTH TIPS

▸ People with recurrent herpes infections should discuss with their physicians the advisability of taking the antiviral drug acyclovir to help ward off recurrent attacks.

---

# What Should You Do if You Suspect You Have Contracted Genital Herpes?

First, see your physician. Treatment with acyclovir or other medication may help during acute outbreaks. While outbreaks resolve naturally, acyclovir may speed healing. Do not engage in unprotected sexual relations from the moment that lesions first appear, or earlier if warning signs precede the outbreak of the infection. You may also wish to ask your physician about the advisability of taking oral acyclovir routinely to reduce the risk of future recurrences. Warm baths, loose fitting clothing, aspirin, and cold, wet compresses may relieve pain during flare-ups of genital herpes. Many herpes sufferers find that subsequent episodes occur less frequently and are milder than initial episodes. Some people go for years or even a lifetime without a recurrence.

---

**HEALTH TIPS**

**If you develop herpes:**

▶ See your physician.

▶ Avoid unprotected genital sex from the moment that sores first appear, or earlier if warning signs precede the infection.

▶ Wearing loose fitting clothing, taking warm baths, using aspirin or other pain relievers, and applying cold, wet compresses may help relieve pain during active episodes.

---

# Coping with the Emotional Consequences of Genital Herpes

The psychological consequences of coping with genital herpes can be more distressing than the physical effects of the disease. The prospects of a lifetime of recurrences and worries about infecting one's sex partners compound the emotional impact of herpes. Feelings of anger, depression, isolation, and shame—even self-perceptions of being tainted, ugly, or dangerous—are common features of the emotional reaction to genital herpes.[27] Herpes sufferers may avoid sexual relations for a long time. Some even restrict their choice of partners to other herpes sufferers.[28] Most herpes sufferers, however, do become sexually active again and learn to lower the risk of infecting their partners by avoiding sexual contact during active episodes. Most also learn to cope with the emotional consequences of herpes.

Attitudes play an important role in a person's ability to cope with herpes. Perceiving herpes as a manageable problem, rather than as a medical disaster or character deficit, may help a person adjust to living with the disease, as this comment from a young woman with herpes suggests:[29]

> Herpes is an inconvenience and a pain, but it's something you
> learn to live with.  I think of it as an imbalance.  Since I know
> it's related to stress, I keep myself in as good physical condition
> as possible and try not to get too upset about it.

Herpes sufferers may also benefit from participating in herpes support groups, which provide opportunities for people to share their feelings about living with herpes with other herpes sufferers and to exchange information about ways of living with the disease.[30]  For information about herpes support groups and other resources that can help you cope with herpes, contact the National Herpes Hotline at (919) 361-8488.

# STD FACT SHEET: HERPES

| WHAT IT IS | A viral infection caused by related strains of the *Herpes simplex* virus. |
|---|---|
| WHAT CAUSES IT | Oral herpes is caused by *herpes simplex virus-type 1 (HSV-1)*; genital herpes is c-aused by *herpes simplex virus-type 2 (HSV-2)*. |
| HOW IT IS TRANSMITTED | Oral, anal, or vaginal sex with an infect-ed partner; oral herpes may also be transmitted by using the same cup, by kissing, or even by sharing a moist towel used by an infected person. Genital con-tact with objects, such as toilet seats, used by an infected person may—but probably infrequently—trans-mit herpes viruses. Genital herpes is almost always spread by sexual contact. |
| SIGNS AND SYMPTOMS | Genital herpes is characterized by the ap-pearance of genital sores or lesions, which may also appear on the thigh or buttocks in both sexes or in the vagina or on the cervix in women. Oral herpes in-volves the appearance of sores or "fever blisters" on the lips, inside the mouth, on the tongue or in the throat. |
| HOW IT IS DIAGNOSED | Clinical examination of sores or lesions; culturing and examination of fluid drawn from a sore |
| HOW IT IS TREATED | Although not a cure, antiviral drugs pro-mote healing of herpetic sores and when given in oral form may reduce the fre-quency of recurrence. |

## CHAPTER ELEVEN

# WHAT YOU SHOULD KNOW ABOUT VIRAL HEPATITIS AND GENITAL WARTS

## *DID YOU KNOW THAT?*

- *A person may be infected with viral hepatitis and not realize it until serious liver problems develop years afterwards.*
- *Hepatitis A may be contracted through contact with infected fecal matter and by eating uncooked shellfish.*
- *A person with viral hepatitis is capable of transmitting the disease to others, even if he or she has no symptoms and is unaware of being infected.*
- *Genital warts have been linked to certain cancers of the reproductive track.*
- *It is estimated that two or three sexually active people in ten in the U.S. are infected with human papilloma virus (HPV), the virus that causes genital warts.*
- *Most people with genital warts have them in areas that cannot be seen, such as on the cervix in women or in the urethra in men.*
- *Women are more susceptible to contracting genital warts than are men.*
- *Although genital warts can be removed by medical treatment, the virus that causes them remains indefinitely within the body.*

Viruses cause a host of illnesses, including the STDs AIDS and herpes. They also cause viral hepatitis and genital warts, both of which may lead to serious medical complications.

# What You Should Know About Viral Hepatitis

Hepatitis is an inflammation of the liver that may be caused by such factors as chronic alcoholism and exposure to toxic materials. **Viral hepatitis** refers to several different types of hepatitis caused by related, but distinct, viruses. The major types are *hepatitis A* (formerly called infectious hepatitis), *hepatitis B* (formerly called serum hepatitis), *hepatitis C* (formerly called hepatitis non-A, non-B), and *hepatitis D*. Approximately 600,000 people become infected with hepatitis each year, about half with hepatitis B.[1]

## What Are the Symptoms of Viral Hepatitis?

Most cases of acute hepatitis occur without symptoms. When symptoms do appear, they often include jaundice (a yellowish discoloration of the skin and body fluids), feelings of weakness and nausea, loss of appetite, abdominal discomfort, whitish bowel movements, and brownish or tea-colored urine.[2]

The symptoms of hepatitis B tend to be more severe and long-lasting than those of hepatitis A. In about 10 percent of cases, hepatitis B can lead to chronic liver disease. Hepatitis C, first discovered in 1975 among hepatitis patients that did not have the hepatitis A or B viruses in their blood,[3] generally gives rise to milder symptoms that may not include jaundice. Yet, despite the mildness of its symptoms, hepatitis C can lead to serious liver diseases, such as cirrhosis or cancer of the liver, both of which can be deadly. About four million people in the U.S. are believed to be infected with hepatitis C. The disease accounts for about 8,000 deaths in the U.S. annually from liver disease— more deaths than from hepatitis A and B combined.[4] The numbers of deaths from hepatitis C are expected to triple by the year 2017. Though people with hepatitis C may be symptom-free at first, half or more eventually develop chronic liver disease. [5]

Hepatitis D—also called *delta hepatitis* or type D hepatitis—is caused by the hepatitis D virus, and occurs only in the presence of hepatitis B. Hepatitis D, which has symptoms similar to those of hepatitis B, can produce severe liver damage that often leads to death.

---

### HEALTH TIPS

▶ Any sign of jaundice, abdominal discomfort, discoloration in the urine or bowel movements, or feelings of weakness or nausea, should be reported to a physician. These may represent early signs of viral hepatitis. Yet most people with acute hepatitis have no symptoms and remain unaware that they are infected unless they later develop serious liver problems.

---

## How is Viral Hepatitis Transmitted?

Hepatitis A and B can both be transmitted sexually, generally through anal sexual contact. Anal intercourse is a frequent means of transmission of hepatitis B, particularly among gay men. It also appears that hepatitis B may be transmitted by heterosexual contact,[6] involving either anal or vaginal intercourse.[7] Hepatitis B may also be spread through transfusion with contaminated blood supplies, by the sharing of contaminated needles among injectable drug users (IDUs), and by contact with contaminated saliva, menstrual blood, nasal mucous, or semen. It is also possible that sharing razors, toothbrushes, or other personal articles used by an infected person may result in transmission of hepatitis B.

The hepatitis A virus appears to be transmitted most frequently through contact with infected fecal matter found in contaminated food or water, and by oral contact with fecal matter, as through oral-anal (licking or mouthing the partner's anus) sexual activity.[8] (It is largely because of the risk of hepatitis A that restaurant employees are required to wash their hands after using the toilet.) Sexual transmission of hepatitis A is believed to be generally limited to cases of oral contact with fecal matter.[9] Ingesting uncooked infested shellfish is also a frequent means of transmission of hepatitis A.[10]

Hepatitis C is primarily transmitted through contact with contaminated blood, most commonly through needle-sharing among injectable drug users and occasionally through transfusions with contaminated blood supplies. Recent evidence shows that hepatitis C may also be transmitted sexually.[11] Like hepatitis B, hepatitis D can be transmitted sexually or through contact with infected blood. A person can transmit the viruses that cause hepatitis, even if he or she is unaware of having any symptoms of the disease.

---

## HEALTH TIPS

**For Preventing the Transmission of Hepatitis:**

▶ Avoid oral-anal sexual contact, since the virus causing hepatitis A and other disease-causing organisms may be transmitted by oral contact with infected fecal matter found in the anus.

▶ Unprotected anal and vaginal intercourse should also be avoided, unless both partners are known to be free of hepatitis and other STDs.

▶ Avoid sharing needles or personal articles such as razors, cuticle scissors, or toothbrushes, with another person.

▶ Avoid eating raw shellfish, since many cases of hepatitis A are traced to eating un c ooked, contaminated shellfish.

---

## How is Viral Hepatitis Diagnosed and Treated?

Viral hepatitis is diagnosed by means of blood tests that detect hepatitis antigens and antibodies. Unfortunately, there is no cure for viral hepatitis. Bed rest and plentiful intake of fluids is usually recommended until the acute stage of the infection subsides, generally in a few weeks. Full recovery may take months. However, the virus may continue to be active in the body, causing permanent liver damage.

There is no cure for viral hepatitis. Some drugs can be helpful, such as alpha interferon, which may help prevent liver damage in people with hepatitis C.[12]

## Hepatitis Vaccines

Vaccines are now available for hepatitis A and B.[13]  The Centers for Disease Control and Prevention (CDC) recommends vaccination of all newborns with the hepatitis B vaccine, as well as teenagers and young adults (check with your doctor). The hepatitis B vaccine also provides protection against hepatitis D, since hepatitis D can only occur if hepatitis B is present. No vaccine for hepatitis C is yet available.

# What You Should Know about Genital Warts

Many people are unaware that warts are caused by viruses. **Genital warts** (also called *venereal warts*) are warts that form on or around the genitals and anus. They are caused by the **human papilloma virus** (HPV).

## What Are the Symptoms of Genital Warts?

Genital warts vary in size and shape. They are similar in appearance to the common plantar warts. They are typically hard and yellow-gray when they form on dry skin. They tend to be pink, soft, and take on cauliflower shapes in moist areas like the lower vagina. Genital warts are generally diagnosed by clinical examination of the affected area.

HPV infection may produce no other symptoms than the warts themselves. However, warts that form on the urethra can cause bleeding or painful discharges.

---

### HEALTH TIPS

▸ HPV infection may produce no other symptoms than the genital warts themselves. Any sign of a wart or growth in the genital area should be reported to a physician.

---

## Where do Genital Warts Usually Appear?

Genital warts usually appear around the genitals and anus within a few months of infection. In men, they usually appear on the penis, foreskin, scrotum, and within the urethra. They usually appear on the external genitalia, along the vaginal wall, and on the cervix in women. They can occur outside the genital area in either gender, such as in the mouth, or on the lips, eyelids, or nipples of the breast, or around the anus or in the rectum.

## Genital Warts Are Linked to Genital Cancers

Genital warts are not dangerous in themselves, but the virus that causes them, HPV, has been linked to cancers of the genital tract, such as cervical and penile cancer.[14]

Cervical cancer is linked to HPV in perhaps as many as 85 to 98 percent of cases.[15] Researchers in one recent study found that women with a history of HPV infection were 11 times more likely than other women to develop cervical cancer during the two-year study period.[16] However, the odds of HPV leading to cervical cancer appear rather slim, as only 13,500 new cases of cervical cancer are reported annually in the United States, as compared to perhaps one million new cases of HPV.[17] (Penile cancers in men are even rarer.) Even so, it would be wise for women to safeguard themselves from HPV-related cervical cancer by limiting their number of sex partners (to reduce their risk of exposure to HPV) and by having regular Pap smears (to detect cervical cancers in their earliest and most treatable stages).[18]

## How Many People Are Affected?

HPV is extremely widespread, perhaps infecting as many as 20 to 30 percent of sexually active people in the United States.[19] One third of college women are believe infected.[20] More than 1 million new cases of HPV infection occur each year in the U.S.

Though genital warts often appear in visible areas of the skin, in perhaps seven of 10 cases they form in areas that cannot be seen, such as on the cervix in women or in the urethra in men.[21] They occur most commonly in young adults age 20 to 24.[22]

## Women Stand a Greater Risk of Developing Genital Warts

Women are more susceptible to HPV infection than men are because cells in the cervix divide swiftly, facilitating the multiplication of HPV.[23] Women who initiate coitus prior to the age of 18 and who have multiple sex partners are especially vulnerable to HPV.[24] Still, men are hardly immune. Researchers report that between 60 and 90 percent of the male sexual partners of infected woman show evidence of genital warts on the penis.[25]

Young, sexually active people are at greatest risk of infection. A recent study of female students treated at a health center at the University of California at Berkeley revealed that nearly half (46%) had contracted HPV.[26] It is estimated that nearly half of the sexually active teenage women in some U.S. cities are infected with HPV.[27]

---

## HEALTH TIPS

▸ Women are at greater risk than men of developing genital warts. Two key ways that women can safeguard themselves from HPV-related cervical cancer is to limit their number of sex partners (to reduce the risk of HPV infection) and to have regular Pap smears (to detect cervical cancer in its earliest and most treatable stage).

---

## How is HPV Transmitted?

In adults, the virus that causes genital warts, HPV, is generally transmitted by skin-to skin contact during vaginal, anal, or oral-genital sex.[28] It can also be transmitted by other forms of contact, such as touching infected towels or clothing. The incubation period between the initial infection and the appearance of warts may vary from a few weeks to a couple of years. Infants can contract the virus from infected mothers as they make their way through the birth canal during delivery.

## How Are Genital Warts Treated?

Freezing the wart (*cryotherapy*) with liquid nitrogen is the treatment of choice for removing the wart.[29] One alternative treatment involves painting the warts with an alcohol-based podophyllin solution over a period of several days, which causes them to dry up and fall off. Podophyllin is not recommended for use with pregnant women or for treatment of warts that form on the cervix.[30] If necessary, the warts may also be treated (by a doctor!) with electrodes (burning) or surgery (by laser or surgical removal). Unfortunately, while the warts themselves may be removed, treatment does not eliminate the virus from the body.[31] There may thus be recurrences.

---

## HEALTH TIPS

► Do *not* attempt to remove genital warts by yourself!  See a doctor.

---

## Prevention of Genital Warts

Unfortunately, no vaccine against HPV exists or appears to be in the offing.  It is possible that the use of latex condoms would help reduce the risk of contracting HPV, although there is a lack of scientific evidence supporting the preventive role of condoms with HPV.[32]  Nor are condoms any use in preventing transmission from areas of the skin they don't cover, such as the scrotum. People with active warts should probably refrain from sexual contact until the warts are removed and the area heals completely.

---

## HEALTH TIPS

► The scientific jury is still out on the role of wearing condoms to prevent the transmission of genital warts.  To play it safe, it is best to avoid all intimate contact with someone during active outbreaks of the infection, until the warts are removed and the affected area heals over.

---

## STD FACT SHEET: VIRAL HEPATITIS AND GENITAL WARTS

| | |
|---|---|
| **WHAT THEY ARE** | Viral hepatitis is a liver disease caused by related strains of hepatitis viruses. Genital warts is a viral infection that produces warts that typically form around the genitals or anus. |
| **WHAT CAUSES THEM** | Hepatitis A, B, C, and D type viruses cause these respective types of viral hepatitis. *Human papilloma virus* (HPV) causes genital warts. |
| **HOW THEY ARE TRANSMITTED** | For viral hepatitis, sexual contact with an infected partner (often involving the anus), transfusion of contaminated blood, or contact with infected fecal matter (especially for hepatitis A); for genital warts, sexual contact with an infected partner, or from infected mother to baby during childbirth. |
| **SIGNS AND SYMPTOMS** | Viral hepatitis may be asymptomatic or involve mild flu-like symptoms or more severe symptoms such as abdominal pain, fever, and the appearance of yellowish (jaundiced) skin and eyes. Genital warts are symptomized by the appearance of warts around the genitals and anus of varying size and shape that may resemble common plantar warts. |
| **HOW THEY ARE DIAGNOSED** | Blood tests are used to diagnose viral hepatitis. Genital warts are diagnosed by clinical inspection of the warts themselves. |
| **HOW THEY ARE TREATED** | No cure exists for either viral hepatitis or genital warts. Viral hepatitis is usually treated with bedrest and, in the case of hepatitis C, possible use of alpha interferon. Genital warts are generally removed by a physician, but the underlying HPV infection remains in the body indefinitely. |

## CHAPTER TWELVE

# WHAT YOU SHOULD KNOW ABOUT OTHER STDS

## *DID YOU KNOW THAT?*

- ► *The body has no natural defense against infestations by certain parasites, such as those that cause "crabs" and scabies.*
- ► *The parasite that causes "crabs" is a tiny insect, not a member of the crab family.*
- ► *You can transfer "crabs" from your genitals to your scalp by touch.*
- ► *You can rid yourself of an infestation of crabs or scabies, only to become reinfected (or infect others) if you fail to wash and dry your clothing, bedding and towels on the hot cycle or have them dry-cleaned.*
- ► *If left untreated, some STDs can lead to a disfiguring skin condition, like that which affected the so-called Elephant Man in the 19th century.*

This chapter focuses on other STDs with which you should be acquainted, including infestations of parasites that are spread by sexual contact and some less common bacterial and viral STDs.

# STDs Caused by "*Yechto*parasites" (Actually *Ecto*parasites)

Ectoparasites are parasitic organisms that live on the outside of the host's body (*Ecto* means "outer"). (Endoparasites, like the protozoan that causes trichomoniasis, *Trichomonas vaginalis*, live within the host's body. *Endo* means "inner."). Ectoparasites are larger than the agents that cause other STDs. Some are large enough to be visible to the naked eye. The body has no effective, natural defense against ectoparasites.[1] Thus, anyone who suffers an infestation should seek appropriate treatment, rather than wait for it to disappear on its own. There are two types of STDs caused by ectoparasites: pediculosis and scabies.

# What You Should Know About Pediculosis ("Crabs")

**Pediculosis** is the name given to an infestation of a parasite whose proper Latin name, *Pthirus pubis* (**pubic lice**), sounds rather too dignified for these bothersome (dare we say ugly?) creatures that are better known as "crabs." Pubic lice are commonly called "crabs" because they are somewhat similar in appearance to crabs when looked at under a microscope (see Figure 12.1). Pubic lice are classified within a family of insects called biting lice. Another member of the family, the human head louse, is an annoying insect that clings to hair on the scalp and often spreads among children at school.

**Figure 12.1 A pubic louse, the organism that causes "crabs"**

```
┌─────────────────────────────────────────────────────────────┐
│                      HEALTH TIPS                              │
├─────────────────────────────────────────────────────────────┤
│  ▸  Pubic lice are *not* of the same family of animals as crabs. They are │
│     somewhat similar in appearance to crabs under the microscope, howev-  │
│     er. They belong to the same family of insects as the human head louse.│
└─────────────────────────────────────────────────────────────┘
```

## What Are the Symptoms of "Crabs"?

Itching, ranging from the mildly irritating to the intolerable,[2] is the most prominent symptom of a pubic lice infestation. The itching is caused by the "crabs" attaching themselves to the pubic hair and piercing the skin to feed on the blood of their hosts. (Yecch!)

## How Long do "Crabs" Survive?

Actually, the life span of these insects is only about one month. However, they are prolific egglayers and may spawn several generations before they die, perpetuating the infestation through subsequent generations.

## How Are "Crabs" Diagnosed?

Diagnosis is usually based on visual examination of the affected area for the presence of the parasite. In the adult stage, pubic lice are large enough to be seen by the naked eye.

## How Are "Crabs" Transmitted?

Pubic lice are generally spread sexually, but they may also be transmitted by contact with an infested towel, bedding, or even toilet seat. While "crabs" can survive for only about 24 hours outside a human host, they may deposit eggs in bedding or towels that can take up to seven days to hatch. Therefore, all bedding, towels, and clothes that have been used by an infested person must be washed and dried on the hot-cycle or dry-cleaned to ensure that they are safe.[3] Sexual contact should be avoided until the infestation is eradicated.

---

**HEALTH TIPS**

▸ You can pick up "crabs" from sexual contact, as well as by contact
with towels, bedding, or even toilet seats used by an infected person.
All bedding and towels used by an infected person must be thoroughly
washed and dried on the hot cycle to remove any remaining lice or any
eggs that may have been deposited.

---

**Can "Crabs" Be Spread to Other Body Parts?** Yes, a person's fingers
may transmit the lice from the genitals to other hair-covered parts of the body,
including the scalp and armpits.

---

**HEALTH TIPS**

▸ Should you develop any signs of an infestation, such as an annoying
genital itch, do not touch other parts of your body, especially hair-
covered areas, after touching your genitals without first washing your
hands thoroughly in soap and water.

---

## How Are "Crabs" Treated?

The good news is that infestations can be treated effectively with a prescription
medication, a 1 percent solution of lindane (brand name *Kwell*), which is available
as a cream, lotion, or shampoo, or with nonprescription medications containing pyre-
thrins or piperonyl butoxide (brand names *RID, Triple X,* and others).[4] *Kwell* is not
recommended for pregnant or lactating women, or for young children.[5] A careful
reexamination of the body is necessary after four to seven days to ensure that all
mature lice and eggs were killed.[6]

# What You Should Know About Scabies

**Scabies** (short for *Sarcoptes scabiei*) is a parasitic infestation caused by a tiny mite.
The mites attach themselves to the base of pubic hair and burrow into the skin, where

they lay eggs and subsist for the duration of their 30-day life span. Scabies are most often found on the hands and wrists, but they may also appear on the genitals, buttocks, armpits and feet.[7] But they do not appear above the neck—thankfully!

## What Are the Symptoms of Scabies?

Like pubic lice, the mites that cause scabies are often found in the genital region, where they cause itching and discomfort. They are also responsible for reddish line-like burrows and sores, welts, or blisters in the skin.

---

### HEALTH TIPS

▸ Itching in the genital area is a sign of a possible parasitic infestation. A scabies infection may be suspected by the appearance of reddish, burrow-type lines and by sores, welts or blisters of the skin in the affected areas.

---

## How is Scabies Transmitted?

Transmission may occur through sexual contact with an infected person, or by touching clothing, bed linen, towels, or other fabrics used by an infected person.

## How is Scabies Diagnosed?

Unlike lice, the mites that cause scabies are too tiny to be seen by the naked eye, but diagnosis can be made on the basis of detecting the mite or its by-products on microscopic examination of scrapings from suspicious-looking burrows.[8]

## How is Scabies Treated?

Scabies, like pubic lice, may be treated effectively with 1 percent lindane (*Kwell*). The entire body from the neck down must be coated with a thin layer of the medication, which should not be washed off for eight hours.[9] But, as noted, lindane should not be used by women who are pregnant or lactating. To avoid reinfection, sex partners and others with close bodily contact should also be treated. Clothing and bed linen used by the infected person must be washed and dried on the hot cycle, or dry-cleaned. As with "crabs," sexual contact should be avoided until the infestation is eliminated.

---

**HEALTH TIPS**

▶ As with an infestation of pubic lice, all bedding and towels used by a person who develops scabies should be washed and dried on the hot cycle, or dry-cleaned.

---

# What You Should Know about Some Less Common STDs

Several types of STDs occur less commonly in the U.S. and Canada than those with which most of us are better acquainted. These less common types of STDs include *chancroid, shigellosis, granuloma inguinale, lymphogranuloma venereum,* and *molluscum contagiosum.*

## What You Should Know about Chancroid

**Chancroid,** or "soft chancre," is caused by the bacterium *Hemophilus ducreyi.* It is more commonly found in the tropics and Eastern nations than in Western countries. The chancroid sore consists of a cluster of small bumps or pimples on the genitals, perineum (the area of skin that lies between the genitals and the anus), or the anus itself. These lesions usually appear within seven days of infection. Within a few days the lesion ruptures, producing an open sore or ulcer. Several ulcers may merge with other ulcers, forming giant ulcers.[10] There is usually an accompanying swelling of a nearby lymph node. In contrast to the syphilis chancre, the chancroid ulcer has a soft rim (hence the name) and is usually quite painful in men. Women frequently do not experience any pain and may be unaware of being infected.[11] The bacterium is typically transmitted through sexual or bodily contact with the lesion or its discharge. Diagnosis is usually confirmed by culturing the bacterium, which is found in pus from the sore, and examining it under a microscope. Antibiotics (erythromycin or ceftriaxone) are usually effective in treating the disease.

---

## HEALTH TIPS

▸ Do not ignore any evidence of a sore, blister, or ulcer in the area of the genitals. Check out any suspicious-looking skin eruption with a physician.

---

## What You Should Know about Shigellosis

**Shigellosis** is caused by the *Shigella* bacterium and is characterized by fever and severe abdominal symptoms, including diarrhea, and inflammation of the large intestine. About 25,000 cases of shigellosis are reported annually.[12] It is often contracted by oral contact with infected fecal material, which may occur as the result of oral-anal sex. It can be treated with antibiotics, such as tetracycline or ampicillin.

---

## HEALTH TIPS

▸ Oral-anal sexual activity may transmit various disease-causing organisms, including hepatitis A (see Chapter 11) and the shigella bacterium.

---

## What You Should Know about Granuloma Inguinale

Like chancroid, **granuloma inguinale** occurs more commonly in tropical regions. It occurs rarely in the U.S. and Canada. It is caused by a bacterium, *Calymmatobacterium granulomatous,* and is not as contagious as many other STDs. Primary symptoms are painless red bumps or sores in the groin area that ulcerate and spread. Like chancroid, it is usually spread by sexual or bodily contact with a lesion or its discharge. Diagnosis is confirmed by microscopic examination of tissue taken from the rim of the sore. The antibiotics tetracycline and streptomycin are effective in treating the disorder. If left untreated, however, the disease may lead to the development of fistulas (holes) in the rectum or bladder, destruction of the tissues or organs that underlie the infection, or scarring of skin tissue that results in **elephantiasis**, a condition that afflicted the so-called "Elephant man" in the nineteenth century.

## What You Should Know about Lymphogranuloma Venereum (LGV)

**Lymphogranuloma venereum** (LGV) is another tropical STD that occurs rarely in Western countries. Some U.S. soldiers returned home from Vietnam with cases of LGV. It is caused by several strains of the *Chlamydia trachomatis* bacterium. A small, painless sore may form on the genitals or on the cervix. The sore may go unnoticed, but a nearby lymph gland in the groin swells and grows tender, and other symptoms mimic those of "the flu": chills, fever, and headache. Other symptoms which may occur include backache, especially in women, and arthritic complaints (painful joints).

If LGV is untreated, complications such as growths and fistulas in the genitals and elephantiasis of the legs and genitals may occur. Diagnosis is made by skin tests and blood tests. The antibiotic doxycycline is the recommended treatment; tetracycline, erythromycin, or sulfisoxazole are alternatives.[13]

---

### HEALTH TIPS

▸ Be aware of any genital lesions, sores, or bumps, *whether or not they are painful or ulcerate.* Such symptoms may indicate the presence of an STD, such as granuloma inguinale or LGV. Although rare in the U.S. and Canada, these diseases can lead to elephantiasis, a disfiguring skin condition, and to other serious consequences. Fortunately, these diseases can be effectively treated with antibiotics.

---

## What You Should Know About Molluscum Contagiosum

**Molluscum contagiosum** is caused by a pox virus that may be spread sexually. The virus causes painless raised bumps or pimples to appear on the genitals, buttocks, thighs, or lower abdomen. These lesions usually appear within two or three months of infection. Most infected people have between 10 and 20 lesions, although the number of lesions may range from 1 to perhaps 100 or more.[14] Pinkish in appearance with a waxy or pearly top, they usually appear about three to six weeks following exposure to the virus. The lesions are generally not associated with serious complications and often disappear on their own within six months. Or they can be treated by squeezing them (like "popping" a blackhead) to exude the whitish center plug. Freezing with liquid nitrogen may also be used to remove the lesions.

However, do not try to treat any lesions on your own.  See your doctor.

---

### HEALTH TIPS

▸ Do not try to remove any skin lesions by squeezing or popping them.
See a doctor.

# STD FACT SHEET:  OTHER STDs

| STD | Cause | Transmission | Symptoms | Diagnosis | Treatment |
|---|---|---|---|---|---|
| Pediculosis ("crabs") | *Pthirus pubis* (pubic lice) | Sexual contact or contact with infested bedding or towels, or toilet seats | Intense itching in genital area | Clinical inspection | Lindane solution (Kwell) or other drugs containing pyrethrins or piperonyl butoxide |
| Scabies | *Sarcoptes scabiei* | Sexual contact or contact with infested bedding, towels or clothing | Intense itching in genital area; appearance of reddish burrows in skin | Clinical inspection | Lindane solution (Kwell) |
| Chancroid | The bacterium, *Hemophilus ducreyi* | Contact with a lesion or its discharge | Cluster of small bumps on the genitals or in or around the anus that ruptures and ulcerates | Clinical inspection, followed by examination of cultured sample of fluid drawn from sores | Antibiotics such as erythromycin or ceftriaxone |
| Shigellosis | The *Shigella* bacterium | Contact with contaminated fecal matter, usually resulting from oral-anal sexual contact | Diarrhea and intense abdominal pain | Microscopic examination of the stools for the presence of the *Shigella* bacterium | Antibiotics such as tetracyline or ampicillin |

| STD | Cause | Transmission | Symptoms | Diagnosis | Treatment |
|---|---|---|---|---|---|
| Granuloma inguinale | The bacterium, *Calymmatobacterium granulomatous* | Contact with a lesion or its discharge | Painless red bumps or sores in the groin that ulcerate and spread | Microscopic examination of a sample of tissue taken from the sore | Antibiotics such as tetracycline or streptomycin |
| Lymphogranuloma venereum (LGV) | Several strains of *Chlamydia trachomatis* | Sexual contact | A small, painless sore on the genitals or on the cervix in women; flu-like symptoms | Blood tests for detection of specific antibodies; culturing of a sample of cells from the affected area | Antibiotics, principally doxycycline |
| Molluscum contagiosum | A poxlike virus | Sexual contact | Painless bumps on genitals, buttocks thighs, or lower abdomen | Clinical inspection; microscopic examination of tissue taken from the lesion | Destruction of each lesion (only by a doctor) by squeezing or freezing them |

## CHAPTER THIRTEEN

# WHAT YOU SHOULD KNOW ABOUT STD PREVENTION: IT'S MORE THAN JUST SAFER SEX

*DID YOU KNOW THAT?*

- *Sexual abstinence does not necessarily involve abstention from all sexual contact.*
- *You have an absolute right not to engage in sexual relations with someone who pressures you for sex, no matter what the other person says.*
- *Masturbation is a convenient and safe sexual outlet.*
- *Though some STDs may be noticeable by the presence of unusual or unpleasant genital odors or by such visible features as bumps, warts, sores, rashes or blisters around the genitals, there is no visible sign or detectable odor to indicate the presence of an HIV infection.*
- *Unprotected anal intercourse is one of the riskiest sexual practices.*
- *The use of latex condoms can help prevent the spread of STDs, but does not entirely eliminate the risk of transmission.*

Up to this point in the book, you should have gotten the message that sexually transmitted diseases are not mere annoyances or inconveniences. They are serious illnesses that can cause pain and suffering, and in some cases can result in serious consequences, including internal damage, infertility, and even death. The good news is that you can reduce, if not eliminate, your chances of contracting an STD by taking reasonable precautions. The precautions discussed here and in Chapter 14 apply to people regardless of their sexual orientation. Prevention is the most effective strategy for controlling the spread of STDs[1], especially viral STDs like herpes and AIDS for which there is no cure or vaccine.[2] Prevention of even one case of an STD may prevent its spread to others, perhaps eventually to you. Though much of the focus of STD prevention involves *safer* sex[1] practices, STD prevention involves more than just safer sex. It also involves avoiding contact with blood products that may be contaminated and with personal articles, such as bedding and towels, that others have used and that may harbor certain STD-causing organisms.

# Hey There! Get with the Program. This Thing is a Killer

What precautions are you taking to avoid AIDS? Let's reverse the question: What lies are you telling yourself to avoid taking the proper precautions? Are you telling yourself that it's safe to have unprotected sex *occasionally* or with only *one* or a *few* partners? Do you draw the conclusion that it's safe to have unprotected sex because your partner is from a good family, is clean-looking, or doesn't sleep around or shoot drugs? Do you believe that "falling in love" first will somehow protect you from AIDS?

The fact is that you won't know whether or not your partner, or partners, are infected by looking at them, inspecting their genitals, or meeting their folks. The AIDS virus is out there in the general population. You may have a lesser risk of infection than a drug addict in an inner-city neighborhood or a sexually active gay male. You may be at lesser risk if you have only one or a few sexual partners, don't engage in needle-sharing, and don't have sex with partners who themselves have engaged in dangerous sexual or injection practices or have had other partners who did. But how can you be certain that your partners are telling the truth about their own past sexual practices, let alone vouch for the sexual practices of their previous sexual partners? The answer is that you can't.

Ask yourself the next time you think about having unprotected sex, "Am I willing to die for sex?" The only way to avoid AIDS is to prevent the virus from finding a way into your body. That's what this chapter is about. But all the

---

[1]    The term *safer* sex rather than *safe* is used because no sexual activity in which there is intimate contact is perfectly safe.

knowledge in the world won't protect unless you decide to get with the program and use it. Since the primary means of transmission of STDs is sexual, this chapter focuses on *safer sex* techniques that sexually active people can use to greatly reduce their risks of contracting or transmitting an STD. First, however, let us consider an alternative to safer sex that is safer than even the safest sex, namely abstinence.

---

### HEALTH TIPS

▸ The most effective strategy for preventing the sexual transmission of STDs is abstinence.

---

# Why Not Abstinence?

Sexual abstinence, or **celibacy**, is a sure way to prevent the sexual transmission of AIDS and other STDs. Although priests and nuns commit themselves to a celibate lifestyle because of their religious vows, few others view lifelong celibacy as an acceptable lifestyle. While you might shudder at the prospects of lifelong celibacy, *temporary* abstinence may be a more attractive alternative, especially if you are single and believe that sexual intimacy should be limited to marriage or to a seriously committed relationship. Practicing temporary abstinence does not commit you to a lifetime without sex. It only means that you have committed yourself to abstaining from intimate sexual contact in the present or the immediate future. Some people are attracted to temporary abstinence because of the opportunity it affords them to focus more of their energies on other interests and goals, such as career interests.

Some people who practice abstinence avoid all sexual activity, even light petting or masturbation. Others draw the line at any sexual activity involving genital contact, while still others engage in manual (hand) genital stimulation with their partners, but not in oral-genital sex or sexual intercourse. As far as preventing STDs is concerned, light petting may be considered safe, so long as sexual contact is limited to touching, rubbing, massaging, or holding one another without directly touching the genitals or engaging in open-mouth kissing in which an exchange of saliva occurs, such as during "French" kissing. Closed-mouth kissing on the lips carries only a remote risk of transmitting STD-causing organisms. However, kissing on the lips should be avoided if one of the partners has active cold sores, which may be a sign of oral herpes.

---

### HEALTH TIPS

▶ Sexual abstinence does not necessarily mean avoidance of all sexual contact. Light petting, for example, can be practiced safely with minimal risks of transmission of STDs.

---

# Setting Limits

People who are committed to sexual abstinence who become romantically involved with others may seek to set limits on sexual intimacy at an early point in their relationships so as to avert later misunderstandings and conflicts. Setting limits involves establishing clear boundaries concerning sexual intimacy. This may avoid a potential conflict in which one partner feels deceived or misled by the other concerning the potential for sexual intimacy.

---

### HEALTH TIPS

▶ People who practice sexual abstinence may help avert misunderstandings by apprizing their partners as early in the relationship as possible.

---

Yet people may feel uncomfortable about limit-setting. Some may feel embarrassed talking about sex with their partners, especially at an early stage of a relationship. Some may not want to risk turning-off or discouraging a partner's interest in them. They may also feel that it is "jumping the gun" to talk about sex before the relationship reaches a stage of physical intimacy. It is certainly true that limit-setting may dampen the enthusiasm of a partner who might like greater sexual intimacy. Yet it may be preferable for both partners to know "where they stand" early in a relationship before becoming further involved in an experience that might prove frustrating or disappointing.

Picking a right time and place to discuss limit-setting is important. Timing is especially important, as you want to ensure that the other person is receptive to discussing the issue. Pick a place in which the two of you are alone and are free of distractions and a time when you are unhurried and unfatigued. If you're not certain whether the time and place are right, sound out your partner by asking permission. You can ask, "There's something on my mind. Is it okay if we talk about it here and now?"

## Broaching the Subject

While our society has become more open toward discussing sexuality, many of us feel embarrassed talking about sex, even when talking to our intimate partners. Talking about sex with partners we are only first getting to know can be even more difficult. To help broach the subject, you might say something like, "I'm a little embarrassed to talk about this, but I think you should know how I feel about sexual intimacy." Chances are that the other person will also feel uncomfortable, but will encourage you to talk and will reciprocate in turn.

## Use "I" Talk

Talk about your attitudes and feelings, rather than preach about what people in general should do in their relationships, or what is "right or wrong" in sexual relationships. For example, rather than saying, "It's not right that people just jump into bed these days," say, "I can't speak for others, but I feel that abstinence is best for me at this point in my life."

## Encourage Your Partner to Disclose Personal Feelings

Encourage your partner to discuss his or her feelings about sexual intimacy. But try not to put your partner on the spot by saying, "Now it's your turn. What do you think?" Rather, say something like, "I know this may be as uncomfortable for you as it is for me, but I am interested to know how you feel, whenever you feel like talking about it." Don't push or press the issue. You might suggest that your partner think about it for a few days. Plan to discuss the issue again at a convenient time in the near future.

## Be Specific

Sexual abstinence means different things to different people. Don't just say that you prefer to remain abstinent and let it go at that. Your partner may have a different concept of abstinence than you do. Clarify what you mean, though not necessarily by giving a detailed description of all the "do's and don'ts." For example, you might say, "While I'm not interested in having intercourse or oral sex at this point in my life, I think we can discover other ways of touching each other that we might like. "

# Saying No to Sexual Pressure

Saying no to sexual pressure is not only an issue for people who adopt sexual abstinence as a goal. No one has a right to pressure you into having sexual relations. While rape is an extreme form of sexual pressure that involves the use of threats or force to coerce an unwilling person into sexual intercourse, sexual pressure may take a more subtle form, such as in the use of persistent verbal pressure or seduction "lines" that aim to manipulate a person into having sexual relations by means of deception or trickery. A survey of 194 male undergraduates in a southeastern university showed that 42 percent admitted to using verbal pressure to coerce an unwilling female partner into sexual activity.[3] In a survey of 325 college undergraduates from a northwestern state university, about one in five of the men admitted to having said things to women they didn't mean to persuade them to engage in sexual intercourse.[4] Women respondents to this survey were more likely than men to have been pressured into sexual relations, with one in four of the women reporting that they had engaged in sexual intercourse although they hadn't wanted to because they had "felt pressured by his continual arguments."[5] About one in fifteen of the *men* reported engaging in sexual intercourse unwillingly as a result of sexual pressure.

The use of verbal pressure and seduction "lines" is so common in dating relationships that they are seldom recognized as forms of sexual coercion. Consider, for example, the "classic" ploy in which the man drives his date home but (conveniently) develops "mechanical problems" on the way there. Alone together on some deserted stretch of road with a stalled car, the woman may feel especially vulnerable to his sexual advances, especially if she does not know the man well and is fearful that he might harm her. He may use her vulnerability to his advantage by gaining sexual favors that she might refuse in a setting in where she felt more secure. Then there is the man who deceives his partner into believing that he really loves her as a ploy to get her into bed. His use of lies and deception is a form of sexual coercion because it involves an attempt to use devious means to curry sexual favors by exploiting his partner's emotional needs. Note, however, that telling someone that you love him or her is not a form of sexual pressure or a seduction line if it is spoken honestly and is not used for purposes of manipulation.[6] The accompanying table gives some examples of sexual pressure lines and some possible replies that may help a person resist them.[7]

---

**TABLE 13.1**
**Saying No To Sexual Pressure Lines[8]**

| | **Sexual Pressure Lines** | **Sample Responses** |
|---|---|---|
| *Lines that Reassure You about the Negative Consequences* | "You can't get pregnant the first time." | "Hey, where did you get your sex education? People can get pregnant any time they have intercourse, even if it's just for one second." |
| | "Don't worry—I'll pull out." (Note that withdrawal before ejaculation may not prevent pregnancy.) | "I know you want to reassure me, but people can get pregnant that way, even without ejaculating. " |
| *Lines that Threaten You with Rejection* | "If you don't have sex, I'll find someone who will." | "I can't believe you are making a threat like this. I'm furious that you would treat love-making like some kind of a job, as if anyone will do." |

| | **Sexual Pressure Lines** | **Sample Responses** |
|---|---|---|
| *Lines that Attempt to Put Down the Refuser* | "You're such a bitch." | "It's hard to believe you want to make love to me and you think calling me names will put me in the mood. I need to leave, now." |
| | "Are you frigid?" | "I resent being called names just because I tell you what I want to do with my body." |
| *Lines that Stress the Beautiful Experience Being Missed* | "Our relationship will grow stronger." | "I know you really would like to get more involved right now. But I need to wait. And lots of people have had their relationship grow stronger without intercourse." |
| *Lines that Imply that One Might Settle for Less* | "I don't want to do anything. I just want to lie next to you." | "The way we're attracted to each other, I don't think that would be a good idea. As much as I care about you, I'd better not spend the night." |
| *Lines to Make You Prove Yourself* | "If you loved me, you would." | "You know I care a lot about you. But I feel very pressured when you try to get me to do something I'm not ready for. It's not fair to me. Please consider my feelings." |

| | Sexual Pressure Lines | Sample Responses |
|---|---|---|
| *Lines that Attempt to be Logical, but Aren't* | "You're my girlfriend—it's your obligation." | "If you think sex is an obligation, we need to think about this relationship right now." (*Watch out* for any such talk—it is very common in abusers and rapists. At best, it's an irrational comment by an immature person.) |
| *Lines that Are Totally Transparent* | "I'll say I love you after we do it." | "Bye, now." (There is no way to deal with a person who would say such a thing.) |

---

# WHAT DO YOU SAY NOW?

## Talking Back to Sexual Pressure

Although the examples of countering statements in Table 13.1 may be helpful in resisting specific pressure lines, you should understand that "saying no" is not a privilege you must earn by out-debating your partner, but a basic right you have in controlling your body. You don't have to come up with reasons to say "no." You don't have to outsmart the other person, or have a quick comeback ready for any argument the other person might raise. You don't have to persuade the other person to see things your way. As Elizabeth Powell points out in her book *Talking Back to Sexual Pressure*, "You don't have to engage in a debate! *You don't have to say anything except 'I don't want to.'* It's your body. It's not as if he's asking you to lend him an object that you own. Your body is not an object."[9]

It's your right to determine how, when, where, and with whom you will share a sexual experience. You may think it is appropriate to offer an explanation in a particular situation. But, as Powell recognizes, "you don't *have* to explain." Your body is not debatable.[10]

Now it's your turn. Though you needn't say anything more than "I just don't want to" if confronted with someone who pressures you for sex, you may prefer to use a more specific response to the particular pressure line that is used. Consider, for example, how you would you respond to the following pressure lines:

1. "Why not? We both need it."

You reply:_____

_____

2. "We'll use protection, it's okay."

You reply:_____

_____

3. "But I thought you loved me!"

You reply:_____

_____

4. "Are you a child or a woman?" or, "Are you a man or a boy?"

You reply:_____

_____

5. "Is something wrong with me?  Aren't you attracted to me?"

You reply:_____

_____

6.  "But I have *needs*!"

You reply:_____

_____

7.  "It's only a natural act. It doesn't mean anything."

You reply:_____

_____

8.  "Grow up!"

You reply:_____

_____

9. "Let me show you how much I love you."

You reply:_____

_____

10. "Now you have me all hot and bothered."

You reply:_____

11. "Don't worry, honey. I'm on the pill," or, "Don't worry, honey. I'll use protection."

You reply:_____

12. "I know you're attracted to me. What's wrong?"

You reply:_____

## Masturbation: A Safer Sex Alternative?

Despite the increased openness about sexuality that has taken place in our society since the 1960s, masturbation largely remains a taboo topic. In the nineteenth and early twentieth centuries, masturbation was widely believed—even by medical authorities—to be physically and mentally harmful. Various physical maladies have been erroneously linked to masturbation at one time or another, including epilepsy, heart attacks, insanity, cancer, sterility, and vaginal warts. Scientists have not found any evidence that masturbation is either physically or psychologically harmful. It does not cause insanity, growth of hair on the hands, warts, or any of the other psychological and physical ills once attributed to the practice. People masturbate for many reasons. Some people masturbate for pleasure or release of sexual tension. Others find that orgasms induced by masturbation can help relieve menstrual cramping, provide temporary relief from anxiety, or help induce sleep. Some people who are concerned about the threat of AIDS and other STDs may use masturbation as a sexual outlet when they are without a regular partner, rather than incur the risks of engaging in casual sexual contacts.

Surveys show that the majority of young adults, even young married adults, masturbate.[11] In a recent study of students in an urban university, 71 percent of the women and 83 percent of the men reported masturbating during the preceding three month period.[12] Sixty-eight percent of the nearly 100,000 *married* women surveyed by *Redbook Magazine* reported masturbating.[13]

Masturbation is a nearly perfectly safe sexual release. It is "nearly" safe because it is possible for certain STDs, such as chlamydia and gonorrhea, to be spread by touch from one part of the body to another. Thus, people who masturbate should avoid touching the eyes or licking the fingers after stimulating the genitals.

Many people, however, view masturbation as unnatural or immoral. People who consider masturbation wrong, harmful, or sinful may experience anxiety or guilt, rather than sexual pleasure, if they masturbate or feel the urge to masturbate. There is nothing wrong with preferring *not* to masturbate. Still, masturbation is a sexual technique that can provide a release of sexual tensions without the risks of STDs incurred by intimate sexual contact with others.

# Safer Sex Techniques

Short of abstinence, the only sure way to prevent the sexual transmission of AIDS and other STDs is to maintain a monogamous relationship with an uninfected partner who does the same.[14] Yet many of us are not involved in lifelong monogamous relationships. Even for those who seek such a relationship, there is always "the first time." Thus, anyone who is sexually active with a partner, or partners, who are not known (*known,* not assumed) to be free of infection should consider him or herself at risk of contracting an STD, even AIDS. For people in this category who plan to engage in sexual activity, which applies to almost all of us at one time or another, here is a set of safer sex guidelines which can reduce the risk of contracting or transmitting STDs, including AIDS.

---

### HEALTH TIPS

▶ If you are sexually active and do not know (know, not guess!) that your partner or partners are free of STDs, you should consider yourself at risk of contracting an STD, even AIDS.

---

## Be Knowledgeable about the Risks

Be aware of the risks of STDs. Many of us try to put the dangers of STDs out of our minds, especially in moments of passion. Make a pact with yourself to refuse to play the dangerous game of pretending that the dangers of STDs do not exist or that you are somehow immune.

## Inspect Yourself and Your Partner

Inspect yourself for a discharge, bumps, rashes, warts, blisters, chancres, sores, lice, or foul odors. Check out any unusual feature with a physician before you engage in sexual activity.

You may be able to work an inspection of your partner into foreplay—a reason for making love the first time with the lights on. In particular, a woman may hold her partner's penis firmly, pulling the loose skin up and down, as if "milking" it. She can then check for a discharge at the penile opening. The man may use his fingers to detect any sign of a disturbing vaginal discharge. Other visible features of STDs include herpes blisters, genital warts, sores, rashes, or bumps, and pubic lice. But note that there is no visible sign of HIV infection. You cannot know whether a person is infected by HIV by looking at the person, or by inspecting the person's genitals.

If you find anything that doesn't look, feel, or smell right, bring it to your partner's attention. Treat any unpleasant odor as a warning sign. Your partner may not be aware of the symptom and may be carrying an infection that is harmful to him or her as well as to you or others. If you notice any suspicious signs, refrain from further sexual contact until your partner has the chance to seek a medical evaluation. Even if your concerns prove groundless, you can resume sexual relations without the uncertainty and anxiety that you would have experienced if you had ignored your concerns.

Of course, your partner may become defensive or hostile if you express a concern that he or she may be carrying an STD. Try to be understanding, and recognize that the social stigma attached to STDs makes it difficult for people to accept the possibility that they may be infected. It may be appropriate to point out that STDs are quite common (among college students, bank officers, military personnel, or . . . fill in the blanks) and that many people are unaware that they carry them.

But for your sake as well as your partner's, if you are not sure that sex is safe, stop. Think carefully about the risks, and seek expert advice.

## Avoid High-Risk Sexual Behaviors

Avoid unprotected vaginal intercourse (intercourse without the use of a latex condom and a spermicide containing nonoxynol-9; see below), unless you are absolutely certain that you and your partner are not carrying an STD.[15] Other high-risk behaviors include unprotected oral-genital activity, insertion of a hand or fist into someone's rectum or vagina ("fisting"), or any activity in which you or your partner would come into contact with the other's blood, semen, or vaginal secretions. Unprotected anal intercourse is one of the riskiest sexual practices. Anal penetration by a penis, or by a partner's hand, carries a heightened risk of infection because tears in the anal lining can provide microorganisms with a convenient port of entry into the bloodstream. Unless you are absolutely sure that you and your partner are free of STDs, especially HIV infection, such activities should be avoided. (It is also advisable to avoid any activity that may be physically injurious.) If you do engage in anal-genital sex and are uncertain as to whether you or your partner are infected, use a latex condom and spermicide. Oral-anal sex, or anilingus (sometimes called *rimming*), should be avoided because of the potential of transmitting microbes between the mouth and the anus.

---

## HEALTH TIPS

► Unprotected anal intercourse is among the riskiest sexual practices and should be avoided unless both partners are known to be free of STDs. Even then, couples should take necessary precautions to avoid injury to sensitive rectal tissue. "Fisting" and oral-anal sex are to be avoided altogether because of the significant risks they pose of infection and of tearing or injury to rectal tissue.

---

## Use Latex Condoms

When used properly, latex (rubber) condoms can help lower, though not eliminate entirely, the risk of contracting or spreading many STDs, including HIV infection and AIDS. Latex (rubber) condoms are particularly effective in preventing the spread of various STDs. They may be even more effective when combined with spermicides containing the ingredient nonoxynol-9, which kills STD-causing microorganisms, including the agents that cause AIDS, chlamydia, syphilis, gonorrhea, and genital herpes.[16] Condoms made from animal membranes ("skins") are less effective as barriers against STD-causing organisms. Even latex condoms are not 100 percent effective in preventing the transmission of the AIDS virus and other STD-causing microorganisms.[17] Condoms (and the people who use them) are fallible. They can break or slip off. They also do not cover all exposed genital areas. Latex condoms may sometimes leak viral particles.[18] Researchers estimate that in regular use, condoms reduce the rate of STD infections by about 50 percent, on the average.[19] Improper use or inconsistent use is a common reason for failures of condoms to prevent STD transmission.[20] Yet, even when used properly, condoms may be of limited or no value against disease-causing organisms that are transmitted externally, such as those causing herpes, genital warts, and ectoparisitic infestations. Still, latex condoms are the best protection against STDs that sexually active people have available. Instructions for using condoms are found in Chapter 14.

## Do Not Rely on Other Contraceptives for Disease Prevention

Most contraceptives provide little or no protection against STDs. Oral contraceptives (the "pill") provide excellent protection against unwanted pregnancy, but do not protect against STDs. Intrauterine devices (IUDs) provide no STD protection. Sterilization (vasectomy for the man; tubal ligation for the woman) offers unequaled protection against pregnancy, but affords no protection against STDs. The diaphragm, a rubber dome which covers the cervix, may provide some protection, especially when it is used along with spermicidal jelly containing nonoxynol-9 placed inside the dome and around the rim. Spermicidal foams, jellies, creams, films, and suppositories containing nonoxynol-9 may also offer some protection against gonorrhea and chlamydia when used alone without a condom. However, they have not been established to provide any protection against HIV infection.[21] Thus, they should not be used in place of the latex condom. Male partners of women using a diaphragm should also wear a latex condom to increase protection against STDs, even if spermicides are used with the diaphragm.

Some "contraceptive" techniques afford little or no protection against either unwanted pregnancies or STDs. In withdrawal (also called *coitus interruptus*), the man withdraws the penis from the vagina before ejaculation. Pregnancy may occur when the man fails to withdraw in time or when viable sperm are found in the pre-ejaculatory fluid that is normally emitted from the penis prior to ejaculation. Withdrawal affords no protection against the transmission of STDs. Fertility-awareness methods (also called *rhythm* methods) include techniques such as the calendar method and the basal-body temperature method, which help the couple predict the period of the woman's maximum fertility each month. The couple then abstains from intercourse during this time to avoid an unwanted pregnancy. Unfortu-nately, fertility-awareness methods are generally ineffective in typical use as a method of contraception and offer no protection against STDs on those occasions in which the couple engages in unprotected sexual intercourse.

## Be Careful in Choosing Your Sex Partners

Choose your partners carefully. The safest course is to avoid sexual contact with someone who is infected with an active STD or is seropositive for HIV, or has engaged in high-risk sexual or drug-use practices and is not known to be seronegative.

---

### HEALTH TIPS

▶  Because sexual partners may carry disease-causing organisms they
   have picked up from previous partners, making love to someone is like
   making love to everyone they've ever slept with in the past.

---

It is not enough to ask your partner about past sexual behavior and drug use (he or she may lie or have lapses of memory). You need to know the person well enough to be able to judge the person's truthfulness. Even then, you cannot be sure that the person is truthful or can completely recall all past sexual experiences, let alone verify the sexual histories of all previous partners. In addition, claims of uninfected HIV status made in the heat of passion are probably lies. To be safe, it is best to abstain or to practice safer sex techniques with any partner who is not known to be free of HIV and other STDs.

---

### HEALTH TIPS

▶  Do not accept at face value what people say about their past sexual
   history or infection status. People may lie (no kidding!) or have lapses
   of memory about past sexual experiences. Besides, how many of us
   can say we're completely knowledgeable about the past sexual histories
   of our sexual partners?

---

Also, avoid sexual contacts with prostitutes, with people who frequent prostitutes, and with people who inject drugs.

## Limit Your Number of Sex Partners

Be selective in your choice of sex partners. Having sex with multiple partners—especially "one-night stands"—increases your risk of eventually having a
sexual contact with an infected person. The greater the number of sexual contacts,
the more likely it is that one or more will be with an infected person. It is similarly
advisable to avoid sex with someone who has had multiple partners.

## Wash Your Genitals Before and After Sex

Washing your genitals before and after sex removes some potentially harmful agents.
Washing together may be incorporated into erotic foreplay. Right after intercourse,
a thorough washing with soap and water may help reduce the risk of infection. Do
not, however, deceive yourself into believing that washing your genitals is an
effective substitute for practicing safer sex. Most STDs are transmitted internally.
Washing is of no avail against them, and only of limited help in preventing infection
of exposed body parts.

---

### HEALTH TIPS

▸ Washing your genitals before, and immediately after, sex may provide
some limited protection against STDs, but is no substitute for practicing safer sex.

---

There may be some limited benefits to women from **douching** right after
coitus. But frequent douching should be avoided since it may change the vaginal
flora and encourage the growth of infectious organisms. Nor is immediate douching
possible for women who use a diaphragm and spermicide that must remain in the
vagina for at least six to eight hours after intercourse. But such women may profit
from washing the external genitals immediately after coitus.

---

### HEALTH TIPS

▸ Preventive measures used after intercourse, such as washing yourself or douching, may provide some limited protection, but should not be used as a substitute for practicing safer sex. Women who use spermicides or a diaphragm should not douche for at least six to eight hours after intercourse, although they may wash their external genitalia.

---

## Engage in Noncoital Activities

Other forms of sexual expression, such as massage, hugging, caressing, mutual masturbation, or rubbing bodies together without vaginal, anal, or oral contact, are low-risk ways of finding sexual pleasure, so long as semen or vaginal fluids do not come into contact with mucous membranes or breaks in the skin.[22] Many sexologists refer to such activities as **outercourse** to distinguish them from sexual intercourse. Sharing sexual fantasies and intimate talk can be highly erotic, as can showering or bathing together, so long as semen or vaginal fluids touch only healthy skin. However, even tiny cuts can allow penetration by viruses and other STD-causing organisms. Vibrators, dildos, and other "sex toys" may also be erotically stimulating and carry a low risk of infection, if they are washed thoroughly with soap and water before use, and between uses by two people. If used for penetration, they should be used gently and with plenty of lubricant to avoid irritating or breaking vaginal or rectal tissues.[23]

---

### HEALTH TIPS

▸ Outercourse (as opposed to intercourse) allows couples to engage in low-risk physical intimacies.

---

## Use Barrier Devices When Practicing Oral Sex

If you do decide to engage in oral sex, use a condom before practicing fellatio and a dental dam to cover the vagina before engaging in cunnilingus.[24]   Condoms are now available in a variety of styles. Dental dams are square pieces of latex rubber that dentists use during oral surgery. Some people cut-open a latex condom to use to cover the vagina during cunnilingus, while others have used household plastic wrap. None of these vaginal barrier methods have been shown to be effective in preventing the transmission of STD-causing organisms, however. Plastic wrap should never be used as a substitute for latex, as it may tear or puncture easily or contain pores large enough to allow microorganisms to pass through. Even latex may be ineffective as a vaginal barrier if it does not fit securely fit over the vagina and allows vaginal secretions to leak around the edges. If you should use a vaginal barrier during cunnilingus, it is best to spread some spermicidal jelly or cream containing nonoxynol-9 on the side that is placed against the vaginal opening to provide added protection.

## Stay Sober

Get in the habit of socializing without using alcohol. If you do use alcohol, drink modestly so that you can maintain a clear head. Excessive alcohol intake can impair your judgment and lead you to take risks you might not otherwise take. Likewise, avoid using drugs that might impair your judgment or ability to think clearly about what you are doing.

## Avoid Sexual Activity When in Doubt

None of the previous safer sex practices guarantees protection. Avoid any sexual activity about which you are in doubt. If you have doubts about a particular sexual activity, discuss your concerns with your health professional or call one of the hot-lines listed in Chapter 15.

# Other Ways of Protecting Yourself from STDs

STD prevention involves more than just safer sex.

## Have Regular Medical Checkups

A sexually active person should be checked for STDs during regular health examinations, preferably twice a year.[25] Many community clinics and family planning centers set their charges according to the patient's ability to pay. Checkups are a small enough investment to make in one's health. Recall that a number of women and men are symptomless carriers of STDs, especially chlamydial infections. Medical checkups can detect disorders that might otherwise go unnoticed. Many physicians advise routine testing of asymptomatic young women for chlamydial infections to prevent the hidden damage that may occur if the infection goes undetected and untreated.[26]

---

### HEALTH TIPS

▶ Some STDs may produce no apparent symptoms and go unnoticed by the infected person. Having a regular medical check-up may reveal the presence of disorders that might otherwise go unrecognized.

---

## Discuss Whether You and Your Partner Should Undergo Testing for STDs Before Initiating Sexual Relations

Some couples reach a mutual agreement to be tested for HIV and other STDs before they initiate sexual relations. (Some people simply insist that their prospective partners be tested before they initiate sexual relations.) But many people resist testing or feel insulted when their partners raise the issue. People usually assume that they are free of STDs if they are symptom-free and have been reasonably "careful" in their choice of partners. The absence of symptoms is no guarantee of freedom from infection, however. Moreover, even "careful" people cannot guarantee that their partners are free of infection.

---

### HEALTH TIPS

▶ STDs happen to the "nicest people." You can't tell whether people carry STDs by examining their social background, physical appearance, or whether they too have been "careful" in their choice of sexual partners.

---

Testing for STDs may reveal the presence of previously undetected infections. Yet people considering testing or insisting that their partners be tested for STDs, especially for HIV, should be prepared for handling the consequences of dealing with positive test results.

## Consult a Physician if You Suspect that You Have Been Exposed to an STD

Consult a physician if you have had unprotected sex with someone you either know or suspect had an STD at the time.[27] Don't wait for symptoms to appear, as symptoms may not appear for years. Early intervention may prevent the damage of an STD spreading to vital body organs. Be aware of any physical changes that may be symptomatic of STDs, and consult a physician when in doubt.

---

### HEALTH TIPS

▶ Don't wait for symptoms to appear if you suspect you have been exposed to an STD. Symptoms may not occur for months or years. Check out any concerns with a physician.

---

## Avoid Other High-Risk Behaviors Which Might Put You In Contact with Someone Else's Bodily Fluids

Avoid contact with bodily substances (blood, semen, vaginal secretions, fecal matter) from other people. Do not share hypodermic needles, razors, cuticle scissors, or other implements that may contain another person's blood. Be careful when handling wet towels, bed linen, or other material that may contain bodily fluids.

# What to Do If You Suspect You Have Contracted an STD

First, contact your personal physician right away. If you can't afford your own physician, call your local health department to see if there is a low-cost clinic available in your community which offers STD treatment. This book may make you better aware of the signs and symptoms of various STDs, but it does not qualify you to diagnose yourself. Second, follow your physician's directions or seek a second medical opinion if you have any questions or doubts about the recommended treatment.

# If You Are Being Treated for an STD...

1. *Take all medication as directed.* Do not skip any doses or combine dosages. If you should inadvertently skip a dose, call your physician for instructions. Discuss any side-effects with your physician. Although the medication may relieve symptoms in a day or two, it may be necessary to continue to take the medication for a week or more to ensure that the infection is completely eliminated.

2. *Understand how to use medication correctly.* Some medication calls for you to abstain from alcohol or to avoid certain foods, like dairy products. Some medications should only be taken before or after meals. Check with your physician or pharmacist to determine the correct way to take any medication.

3. *Abstain from sexual contact during an active infection.* Although using a latex condom combined with a spermicidal agent containing nonoxynol-9 may provide some protection against disease transmission, it is best to abstain from sexual activity until the infection clears. Consult your physician regarding the recommended duration of abstinence.

4. *Contact sexual partners who may have infected you or whom you may have infected.* Suggest that they seek a medical evaluation to see if they too are infected. It is possible that they are infected and are unaware of the problem because of a lack of symptoms. Remember that STDs can cause internal damage and be passed along to others, even if they do not produce any noticeable symptoms.

5. *Return for follow-up visits if your physician instructs you to do so.* Although the symptoms of the infection may be relieved after a few days, return for any requested follow-up visits to ensure that you are free of the infection. Your physician may also request that your partner be evaluated so that the two of you do not bounce the STD back and forth.

6. *If you have a continuing STD, such as herpes or HIV, share the information with your partner or partners. Don't keep it a secret.* Your sexual partners have a right to know about your infection status before initiating or resuming sexual relations. Make sure that both you and they understand the risks and the necessary precautions that may need to be taken. If you have any doubts about the safety of engaging in sexual relations, consult your physician before engaging in any sexual contact. It may be helpful for both you and your partner to sit down with your physician and discuss the risks you face and the concerns that either of you may have. Unfounded fears and concerns can often be allayed by receiving corrective information.

# What to Do if You Test Positive for HIV

People who learn that they are HIV-infected may experience a range of emotions, including anger, depression, fear, and anxiety. Some people, however, react with an attitude of indifference and attempt to deny the seriousness of the problem. Whatever your initial reaction to learning that you are HIV-positive, do not give up hope. Many people today are living full and satisfying lives for many years after HIV infection. Treatment approaches under investigation offer hope of prolonging a healthy state in people with asymptomatic HIV infection or delaying or possibly averting the development of full-blown AIDS, or increasing survival of of people with AIDS.

Upon learning that you are HIV-infected, consult your physician or health professional. Ask to have the Western blot test performed to confirm the initial positive finding. About one percent of initial *seropositive* blood tests are incorrect. Assuming that the diagnosis is correct, discuss with a health professional the steps you need to take to bolster your health. Many of the suggestions actually mirror the advice that health authorities offer to people in general: get plenty of sleep, adopt a well-balanced diet, learn to avoid stress or more effectively handle the stress you cannot avoid. Now, however, you will need to pay even more serious attention to your general health to help offset the damage that HIV can do to your immune system. Discuss with your health provider the treatment options that are available, including the use of antiviral drugs and other drugs intended to prevent the development of opportunistic diseases that may arise from damage to your immune system caused by HIV. Learn as much as you can about HIV infection and AIDS and take an active role in your care. Become better aware of the actions and side-effects of any drugs that you may be prescribed. Don't be afraid to ask questions or to seek a second opinion. But most importantly, begin treatment as early as possible. The earlier that treatment of HIV infection begins, the better your long-term prospects.

Notify any of your sexual partners whom you possibly could have infected, or ask your health provider to do so. Try not to alarm them (the risks of infection are rather small from a single or a few encounters), but suggest that they think seriously about being tested for HIV.

The majority of people with HIV and AIDS experience emotional problems, including problems relating to anxiety, fear of death, depression and a sense of hopelessness and alienation. Don't try to go it alone. Seek support from your family and friends. There may be an initial tendency to retreat from others, as you may fear that casual contact might expose others to a risk of infection. Others too, may be initially rejecting or afraid, perhaps because they too harbor unreasonable fears about the risks of HIV transmission. Educate yourself about HIV/AIDS and help others to overcome prejudices by becoming better aware of how HIV is, and is not, transmitted. Recall that HIV is not transmitted by casual contact with others, or by sharing the same house, bathroom, or workplace. Seek other sources of support, such as religious leaders, therapists, and other people with HIV. Contact support groups in your area for people with HIV, such as your local AIDS support organization. If you don't know how to locate a support group in your area, call the CDC National AIDS Hotline at 1-800-342-AIDS. You needn't give your name or identify yourself.

## CHAPTER FOURTEEN

# WHAT YOU SHOULD KNOW ABOUT CONDOMS AND HOW TO USE THEM

## *DID YOU KNOW THAT?*

▸ *The use of penile sheaths dates back to ancient Egypt, if not earlier.*

▸ *Condoms made from animal membranes (so-called naturals or skins) have pores that can allow tiny viruses to pass through, including HIV.*

▸ *Spermicides containing nonoxynol-9 not only kill sperm, but also many disease-causing organisms.*

▸ *You should not lubricate a condom with petroleum jelly.*

▸ *Used condoms should not be flushed down the toilet.*

▸ *You should never test the strength of a condom by stretching it or inflating it.*

▸ *You should practice safer sex each and every time you engage in sex.*

It is ironic that in this technologically-advanced age in which we live, the best means of preventing the sexual transmission of STDs is a decidedly low-tech device: the condom. While improvements in materials and manufacturing standards have been made over the years, the condom has remained essentially the same since rubber was first galvanized by Charles Goodyear in 1843. The history of the condom can be traced back much further in time, at least as far back as ancient Egypt, where penile sheaths were worn as decorative covers as early as 1,350 B.C.[1]

Penile sheaths made of linen were first described in European writings in 1564 by the Italian anatomist Fallopius (from whose name the term fallopian tube is derived). They were used, without success, as a barrier against syphilis. The term condom was not used to describe penile sheaths until the 18th century, when sheaths made of animal intestines became popular as a means of preventing sexually transmitted diseases and unwanted pregnancies. Condoms made of rubber (hence the slang "rubbers") came into use shortly after Goodyear developed his galvanization process.

Condoms lost popularity as a contraceptive device with the advent of the pill and the IUD. However, condoms have been making a comeback because they (especially the latex type) can help prevent the spread of the AIDS virus and other STDs, and to a lesser extent because of concerns about side effects of the pill and the IUD. Today, latex rubber condoms come in various styles and are available without a prescription from your neighborhood drug store. They are even sold in vending machines in some college dormitories. With the increased threat of STDs in recent years, condoms have acquired a certain trendiness, as witnessed by the range of "designer" colors that are now available. Speciality stores featuring a wide assortment of condoms have sprouted-up, like "Condomania" in New York City. There are even drive-by condom kiosks in some shopping centers, where shoppers can pick up supplies without even leaving their cars. Having succeeded in dominating the electronics industry, the Japanese are now making significant inroads in manufacturing extremely thin latex condoms for the U.S. market, which undoubtedly will increase our ever-burgeoning trade deficit with Japan.

# What is a Condom?

The condom is a cylindrical sheath that is unrolled onto the erect penis by the man or his partner. They are used for contraceptive and **prophylactic** purposes. As a contraceptive, condoms serve as a mechanical barrier to prevent the passage of sperm from the man to the woman. Conception cannot occur without the uniting of a sperm cell and a ripened egg cell.

As a prophylactic, condoms prevent disease-carrying microorganisms from being transmitted from the man's penis to the woman. They can also help prevent infectious microorganisms found in the woman's vaginal fluids from finding their way into the man's body through the penile opening or through tiny openings in the skin of the penis. Latex condoms can help prevent the transmission of various STD-causing organisms, including those responsible for AIDS, hepatitis, and herpes.

Latex condoms are even more effective in preventing disease transmission when they are used along with spermicides (chemicals that kill sperm) that contain the ingredient nonoxynol-9. This ingredient not only deactivates sperm but also kills various STD-causing microorganisms, including those that cause AIDS, chlamydia, syphilis, gonorrhea, and herpes.[2] The spermicide should be spread on the outside of the condom and inserted within the vagina by using an applicator. Some condoms already come equipped with a spermicide containing nonoxynol-9 as a lubricant, but use of additional spermicide applied within the vagina may yet offer greater protection.

---

### HEALTH TIPS

▸ Sexually active men and women should carry latex condoms with them and insist that they be used during any sexual act involving genital contact.[3]

---

## Advantages and Disadvantages of the Condom

Condoms have a number of advantages as prophylactics and contraceptives. They are readily available for use as needed and don't require prior medical consultation. They can be purchased without a prescription and do not require a special fitting. They can remain in sealed packages until they are needed. Since they are simple mechanical barriers, condoms do not affect production of hormones or fertility. Women ovulate normally; men produce sperm and ejaculate normally. Condoms are almost entirely free of side effects (an advantage reported by 70 percent of female respondents to a *Consumer Reports* survey[4]).

The major disadvantage of the condom as a form of birth control is that is associated with a very high failure rate in typical use. The failure rate among typical condom users is estimated to be about 10 percent per year.[5] That is, one in ten typical users will get pregnant within a year. Most pregnancies are due to the failure to use the condom reliably or correctly. Yet, when the condom is used correctly, and when it is used in combination with spermicides, the failure rate drops to 2 or 3 percent per year, which rivals the effectiveness of the birth control pill.[6]

One limitation of the condom as a prophylactic is that it sometimes fails to prevent disease transmission. Condoms, and the people that use them, aren't perfect. Condoms sometimes slip off or tear, allowing disease-causing organisms to pass from partner to partner. But most failures of the condom as a prophylactic occur when couples fail to use them correctly or consistently (that means every time they have sex), or fail to use them at all.

Another limitation of the condom as a prophylactic is that it does not provide protection against all STDs.[7] Even when used properly, latex condoms offer no protection against pubic lice, which cling to hair-covered areas of the genitals left uncovered by the condom. Genital herpes is spread by contact with infected areas of the skin or mucosal lining of the vagina. Condoms are ineffective in preventing the spread of herpes unless the sores are limited to the penis or lining of the vagina.

Other disadvantages of using condoms (discussed below) include some loss of sexual sensation, especially for the man, and the need to interrupt the sexual act in order to put the condom on the penis following erection. Despite its limitations, the condom represents the best means available for sexually active people to protect themselves and their partners from the spread of STDs.

# What Are the Different Types of Condoms?

Natural condoms (called "skins" or "lambskins") are made from the intestinal membranes of animals. While they are more expensive than latex rubber condoms ("rubbers") and allow greater sexual sensation, natural condoms, as noted in Chapter 13, do not protect as well against STDs as the latex type. Natural condoms have pores that may be large enough to let HIV and other viruses, such as the one that causes hepatitis B, slip through.[8] Latex condoms prevent the passage of these and other microorganisms.

---

### HEALTH TIPS

▶ Since only latex condoms are demonstrated to be effective against the tiny AIDS virus, do not use natural-skin condoms in place of latex condoms if you are concerned about protecting yourself and your partner from AIDS and other STDs.

---

Condoms that provide protection against STDs carry a claim of disease prevention on the package. Condoms that carry this claim must meet government standards. New to the market, the female condom (a lubricated polyurethane sheath placed in the vagina before sexual contact; brand name *Reality)* has not yet been shown to effective in preventing HIV infection and other STDs. [9]

---

## HEALTH TIPS

▸  Don't assume that the more expensive condoms provide better protection against STDs.  Only condoms that carry a claim on the package
   that they help prevent the transmission of sexually transmitted diseases
   are effective in preventing STDs.

---

       Condoms are either plain-tipped or have nipples or reservoirs at the tip to collect semen following ejaculation. Figure 14.1 shows the correct way for putting on a condom.

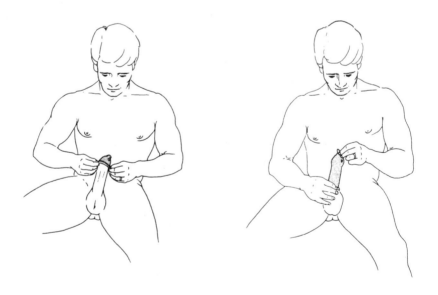

**Fig. 14.1  Putting on the Condom.**

# How to Use a Condom

The condom should be rolled onto the penis by the man or his partner following erection, and before intercourse or any contact between the penis and the partner's genitals. Immediately after use, the condom should be checked for tears before it is discarded. If a condom falls off or slips off during use, or if any tears are found, a spermicide should be used immediately to help prevent conception and provide greater protection against STDs. Additional instructions for condom use are given in Table 14.1.

---

**TABLE 14.1**
## How To Use a Condom[10]

- Use a condom each and every time you have intercourse. Inexperienced users should practice putting on a condom before they have the occasion to use one with a partner.
- Handle the condom carefully, making sure not to damage it with your fingernail, teeth, or other sharp objects.
- Place the condom on the erect penis before it touches the vulva.
- For uncircumcised men, pull back the foreskin before putting on the condom.
- If you use a spermicide, place some inside the tip of the condom before placing the condom on the penis. You may also wish to use additional spermicide applied by an applicator inside the vagina to provide extra protection, especially in the event of breakage of the condom.
- Do not pull the condom tightly against the tip of the penis.
- For a condom without a reservoir tip, leave a small empty space—about a half-inch—at the end of the condom to hold semen. Some condoms come equipped with a reservoir (nipple) tip that will hold semen.
- Unroll the condom all the way to the bottom of the penis.
- If the condom breaks during intercourse, withdraw the penis immediately and put on a new condom and use more spermicide.
- After ejaculation, carefully withdraw the penis while it is still erect.
- Hold onto the rim of the condom as the penis is withdrawn to prevent the condom from slipping off.

● Remove the condom carefully from the penis, making sure that semen doesn't leak out.
● Check the removed condom for evidence of any tears or cracks. If any are found, immediately apply a spermicide containing nonoxynol-9 directly to the penis and within the woman's vagina. Wrap the used condom in a tissue and discard it in the garbage. Do not flush it down the toilet, as condoms may cause problems in the sewers. Wash your hands thoroughly with soap and water.

---

## Avoiding Condom Failure: What Not to Do with Condoms[11]

It should be no surprise that a condom works best when it is used correctly. Mistakes in using condoms correctly can result in both STD transmission and unwanted pregnancies. Among the most common mistakes are the following:

▶ Rolling the condom on the flaccid (limp) penis
▶ Not putting on the condom until after intercourse has begun.
▶ After ejaculation, allowing the penis to become fully limp before withdrawing it from the vagina.
▶ Using a lubricant like petroleum jelly (such as "Vaseline") or other oil-based lubricant (oil-based lubricants can break down the latex material).
▶ Failing to roll the condom all the way down the penis.
▶ Failing to leave a reservoir tip at the end of a plain-tipped condom.
And most commonly:
▶ Failing to use a condom at all or using it only occasionally.

Here are some things you should *never* do with a condom:

▶ Never use teeth, scissors, or sharp nails to open a package of condoms. Open the condom package carefully to avoid tearing or puncturing the condom.
▶ Never test a condom by inflating it or stretching it.
▶ Never use an oil-based lubricant, such as petroleum jelly (like "Vaseline"), cold cream, baby oil or lotion, mineral oil, vegetable oil, "Crisco", hand or body lotions, and most skin creams. If a lubricant is desired, use a water-based lubricant such as contraceptive jelly or "K-Y" jelly. Do not use saliva as a lubricant because it may contain infectious organisms, such as viruses.

- Never use a damaged condom. Condoms that are sticky or brittle or otherwise appear damaged should not be used. A condom that sticks to itself or feels gummy is damaged and should not be used.
- Never use a condom after the expiration date (if any). The expiration date is not the manufacturing (mfg) date.
- Never use a condom if the sealed packet containing the condom is damaged, or cracked or brittle, as the condom itself may be damaged or defective.
- Do not open the sealed packet until you are ready to use the condom. A condom contained in a packet which has been opened can become dry and brittle within a few hours, causing it to tear more easily. The box that contains the condom packets, however, may be opened at any time.
- Never use the same condom twice. Use a new condom each time you have intercourse. Also, use a new condom if you switch the site of intercourse, such as from the vagina to the anus, or from the anus to the mouth, during a single sexual act.
- Do not store condoms where they could be exposed to extreme heat, such as the glove compartment of a car. Condoms can be eroded by exposure to body heat or other sources of heat. Condoms should be stored in a cool, dry place, such as a medicine cabinet or dresser drawer.
- If you want to carry a condom with you, place it in a loose jacket pocket or purse, not in your pant's pocket or in a wallet held in your pant's pocket, where it might be exposed to body heat.
- Never buy condoms from vending machines that are exposed to extreme heat or placed in direct sunlight.

---

### HEALTH TIPS

- Do not purchase a condom from a vending machine that is exposed to extreme heat or placed in direct sunlight.

# Reasons that People Give for Not Using Condoms

Latex condoms are effective barriers against both sperm and STD-causing organisms. Yet, many sexually active people resist using condoms. The question is, why? Here, let us discuss some of the most common reasons given by sexually active people for not using condoms:

1. *I just didn't have one available.* A condom can't protect you if you don't have one available when you engage in sex. Let's face it. In moments of passion, you might not have the presence of mind to stop what you are doing and dash off to the local 24-hour drug store (assuming one is available) to pick up a package of condoms. You should either carry a condom with you, or make sure one is handy, whenever you might possibly engage in sexual activity. Carrying a condom with you does not mean that you are planning to, or committed to, engage in sex on a particular occasion. It only means that you are prepared to engage in sex, should you decide to do so at the time. On the other hand, not carrying a condom is no guarantee that you won't be swept up in a moment's passion and cast prudence aside by engaging in unprotected sex.

If you do bring a condom along on a date it is best to carry it loosely in a jacket pocket or handbag, and for no more than a few hours at a time, to prevent it from becoming damaged by exposure to excessive heat.

2. *Condoms interrupt the sexual act.* Using a condom may interrupt the sex act since it can only be used once the man achieves an erection. However, putting a condom on the penis takes only a few seconds, a mere momentary lull in the sexual act, and one well worth the inconvenience for the protection and peace of mind that it affords. Moreover, many couples find it to be erotically stimulating to share the experience of putting on the condom, or having the man's partner place the condom on his penis.

3. *They cost too much.* Condoms do cost money, but generally less than a dollar each for the latex variety. Ask yourself, is it not worth some pocket change to prevent getting a deadly disease like AIDS or incurring an unwanted pregnancy?

4. *They dull sexual sensations.* Condoms may lead to a loss of some sexual sensation, especially for men. Natural condoms may provide more natural sensations, but are less effective barriers against STD-causing organisms, especially HIV. Even the latex condom permits enough sexual sensation that most men find the experience to be enjoyable. Men may find that sensations feel more natural if they place some non-oil-based lubricant in the tip of the condom before rolling it on the penis, and some more on the outside of the condom for added lubrication. Again, using the condom is something of a trade-off: A loss of some sexual sensation for

the benefit of protection against STDs, especially AIDS. The peace of mind that comes from protecting oneself and one's partner from STDs and unwanted pregnancies may more than offset the small loss of sexual sensation. Some men (and their partners) find an added benefit of increased ejaculatory control, allowing them to prolong intercourse.

5. *It's not manly to use them.* Our notion of manliness incorporates a sense of responsibility to oneself and others. It is manly, and *womanly*, to take precautions to reduce the risk of transmitting or contracting a deadly disease like AIDS.

6. *I'm too embarrassed to buy them.* It is quite normal to feel embarrassed when buying condoms from your local store, especially for the first couple of times. Despite the increased sexual openness in our society in recent years, we are still rather prudish about sexual matters. Tell yourself that it is quite normal to feel embarrassed. Tell yourself that if the store didn't expect people to buy condoms, they wouldn't stock them, let alone display them so prominently. Tell yourself that embarrassment is likely to ease after the first few times and that purchasing condoms will begin to feel as natural as purchasing any other "personal" product, such as tampons or jock itch spray. No one will ask you any questions about your sex life or put you on the spot to explain why you need them. Still, you may feel more comfortable buying condoms from a store in which you don't usually shop or from a vending machine. Feelings of embarrassment may be uncomfortable but they shouldn't prevent you from making a purchase that could save your life.

7. *I know my partner is safe.* Many people believe that they can somehow tell whether or not their partner is infected with HIV. Wrong. HIV may produce no telltale signs or symptoms during the first few years of infection. A person can be well-groomed and look like the boy or girl next-door and still carry HIV. A person can be physically healthy in every obvious respect, play varsity sports, or run several miles a day, and still carry HIV. A person can have had but one or two sexual partners in the past and still carry HIV. Unless you know that you and your partner are free of HIV, you should consider yourself and your partner to be at risk of becoming infected by engaging in unprotected sex.

Nor is it safe to practice safer sex for a few times and then begin to "wing it" without protection. Don't assume that after a few (or even many) times you'll know whether or not your partner is infected. Unfortunately, HIV can linger undiscovered in the body for years before it reveals itself in the form of obvious symptoms. Nor does "falling in love" make it safe to practice unprotected sex. As the classic Tina Turner song goes, "What's love got to do with it?" HIV respects no boundaries, not even the invisible boundary of love.

8. *My partner didn't have one with him (her).* The responsibility for practicing safer sex rests with both partners. Men and women should carry condoms with them whenever they might engage in sexual activity.

---

### HEALTH TIPS

▶ In order to reduce your risk of contracting an STD, you should use a latex condom and a spermicide containing nonoxynol-9 every time you engage in sex, not just the first time, not just occasionally, not just if you or your partner "feel like it," but every time.

---

9. *I was taught that it is immoral to use artificial contraceptives like condoms.* Many people have been exposed to religious teachings that regard the use of artificial contraception, including the use of condoms, as immoral. People who become sexually active need to weigh religious prohibitions against the risks incurred by practicing unprotected sex.

# What To Say and Do if Your Partner Says No

What should you say and do if you desire to use a condom and your partner says no? Here are some suggestions: First, don't use your partner's refusal as a reason for launching into a tirade about what an inconsiderate, insensitive, uncaring or irresponsible person he (most refusers are men) is. Nor should you suspend your own best judgment and go along with your partner's wishes "just to be nice." Rather, calmly and clearly explain that it's too dangerous to engage in unprotected sex with the threat of AIDS and other STDs hanging around. If your partner claims that you don't have anything to worry about, point out that people can carry HIV for years without realizing it. Mention that using a condom protects you both, and that you care enough about your partner to want to protect the two of you.

What if your partner still refuses to use a condom? Here are some some suggestions. Again, don't use this as an occasion for attacking your partner for being uncaring or insensitive. Such attacks are likely to provoke a counterattack and can disrupt, or even sever, a relationship. Stick to the issue, which most simply is, *no condom, no sex.* Tell your partner that until you resolve this issue, no sexual activity below the belt will occur. Other activities are okay, such as hugging, dry kissing, massage of nongenital areas, but contact involving the vagina, penis, or anus, is not. If your partner continues to insist upon engaging in unprotected sex, you should reassess your relationship and think seriously about finding another partner. Ask yourself whether your partner truly cares about you if he or she is so unwilling to protect you both by practicing safer sex.

Table 14.2 contains some sample replies which you may find helpful if your partner refuses to use a condom. Don't be concerned about the precise wording. Use your own words in communicating with your partner. But most importantly, bear in mind that no one has the right to make you do anything with your body that you are not comfortable doing, including engaging in unprotected sex. You have an absolute right over how you use your body, including the right to say no to sex.

**TABLE 14.2**
**What to Say When Your Partner Says No**

| Your Partner Says | You Say |
| --- | --- |
| "If you really loved me, you wouldn't make me wear a condom." | "Love works two ways. If you cared enough about me you'd want to protect us both by using a condom." |
| "It just doesn't feel natural." | "I know what you mean, but it's just not worth risking our lives to do it without one." |
| "Why don't we just do it just this time without a condom." | "We can't take chances with a virus like HIV. You can get it from having sex just one time." |
| "Relax, everybody does it." | "Because other people are foolish is no reason to follow them." |
| "No one I ever slept with had HIV." | "How can you know for sure? The virus may linger in a person's body for years before symptoms appear." |

"I'm not gay or a drug-user. What do you have to worry about?"

"It's not just gays and drug users who get HIV. Sleeping with someone today is like sleeping with everyone they've ever slept with. Who knows what's floating around in your bloodstream or mine? It's best to play it safe."

"I hear condoms aren't as effective as they claim."

"They may not be perfect, but they're a lot better than using nothing at all."

"I'd like to use one, but I forgot to bring one along."

"Fortunately, I didn't forget. Let me get one."

"I don't really know how to use it correctly."

"It's really easy. Let me show you how. It can be kind of sexy."

"I'll lose my erection if I use one."

"Leave it to me. I'll help make sure you get it back."

## CHAPTER FIFTEEN

# HOTLINES TO SEXUAL HEALTH

D o you have questions about the signs and symptoms of STDs that haven't been answered in this book or in your other reading or discussions?  Do you need assistance in coping with an STD?  Or would you just like to talk to someone about your concerns or fears?  A number of organizations have established telephone hotlines that provide anonymous callers with information and referrals to helper organizations in their areas.  Some organizations publish newsletters and other material to help sufferers of particular diseases cope more effectively.

## National Toll-free Hotlines for Information about AIDS and Other STDs

These hotlines provide information about AIDS and other STDs, as well as referral sources.  You needn't give your name or identify yourself to obtain information.

National AIDS Hotline, Centers for Disease Control AIDS Hotline:
    1-800-342-AIDS.  Information and referral resources nationwide,
    24 hours a day.

National STD Hotline: 1-800-227-8922.  A hotline sponsored by the American
    Social Health Association that dispenses information about STD symptoms
    and refers callers to local STD clinics that provide confidential, minimum
    or no cost treatment.

Spanish AIDS/SIDA Hotline: 1-800-344-7432
AIDS Hotline for Teens: 1-800-234-TEEN
AIDS Hotline for the Hearing Impaired: 1-800-243-7889
Canadian Toll-Free Hotline (Toll free in Canada): AIDS Committee of
    Toronto: 1-800-267-6600

# Where to Call or Write for Additional Information About AIDS and Other STDs

National AIDS Information Clearinghouse: (800) 458-5231
Education Database Distribution
1600 Research Blvd.
Rockville, MD 20850

American Red Cross: (202) 737-8300
AIDS Education Office: (703)206-7180
8111 Gatehouse Road, 6th Floor
Falls Church, VA 22042-1203

National Initiative for AIDS and HIV Prevention among Adolescents
1025 Vermont Avenue, NW
Suite 210
Washington, DC 20005

Teens Teaching AIDS Prevention
3030 Walnut Street
Kansas City, MO 64108

National Association of People with AIDS: (202) 898-0414
1413 K St., NW, 7th Floor
Washington, DC 20005

The Henry Nichols Foundation
P.O. Box 621
Cooperstown, NY 13326

AIDS National Interfaith Network: (202) 546-0807
110 Maryland Avenue NE
Washington, DC 20002

Sexuality Information and Education Council of the United States (SIECUS)
(212) 819-9770
130 West 42nd St., Suite 2500
New York, NY 10036
A national clearing house for information about sexuality, SIECUS provides bibliographies on topics involving human sexuality and can direct callers to agencies, hotlines and other resources to help them with their personal questions regarding AIDS or other sexual topics.

Gay and Lesbian Medical Association: (415) 255-4547
459 Fulton Street
San Francisco, CA 94102

# Where to Obtain Help or Information about Herpes

National Herpes Hotline: (919) 361-8488

*The Helper* is a newsletter published by HELP (Herpetics Engaged in Living Productively), an organization that helps herpes sufferers cope with the disease. For copies of the newsletter, and for the address of the HELP chapter closest to you, either call the National STD Hotline (1-800-227-8922) or write to HELP, Herpes Resource Center, P.O. Box 13827, Research Triangle Park, NC 27709.

Herpes Resource Center
Box 100
Palo Alto, CA 94302

# GLOSSARY

**Acquired immunodeficiency syndrome**
A condition caused by the *human immunodeficiency virus* (HIV) and characterized by destruction of the immune system so that the body is stripped of its ability to fend off life-threatening diseases.

**Antibodies** Specialized proteins produced by the white blood cells of the immune system in response to disease organisms and other toxic substances.

**Antigens** Proteins, toxins, or other substances to which the body reacts by producing antibodies. (Combined word formed from *anti*body *gen*erator.)

**Bacteria** Plural of bacterium, a class of one-celled microorganisms that have no chlorophyll and can give rise to many illnesses. (From the Greek *baktron*, meaning "stick" and referring to the fact that many bacteria are rod-shaped.)

**Celibacy** (1) Abstention from sexual intercourse. (2) A state of being unmarried.

**Cervicitis** Inflammation of the cervix.

**Chancre** A sore or ulcer.

**Chancroid** A bacterial STD caused by the *Hemophilus ducreyi* bacterium. Also called *soft chancre*.

**Congenital syphilis** A syphilis infection that is present at birth.

**Douching** Application of a jet of liquid to the vaginal as a rinse. (From the

Italian *doccia*, meaning "shower bath".)

**Elephantiasis** A disease characterized by enlargement of parts of the body, especially the legs and genitals, and by hardening and ulceration of the surrounding skin. (From the Greek *elephas*, meaning "elephant," referring to the resemblance of the affected skin areas to elephant hide.)

**Fallopian tubes** Tubes that extend from the upper uterus toward the ovaries and conduct ova to the uterus. (After the Italian anatomist Gabriel Fallopio, who is credited with their discovery.)

**General paresis** A progressive form of mental illness caused by neurosyphilis and characterized by gross confusion.

**Genital warts** An STD that is caused by the *human papilloma virus* and takes the form of warts that appear around the genitals and anus.

**Genital herpes** An STD caused by the Herpes simplex virus type 2 and characterized by painful shallow sores and blisters on the genitals.

**Gonococcal ophthalmia neonatorum**
A gonorrheal infection of the eyes of newborn children who contract the disease by passing through an infected birth canal. (From the Greek *opthalmos*, meaning "eye.")

**Granuloma inguinale** A tropical STD caused by the *Calymmatobacterium granulomatous* bacterium.

**Herpes simplex virus type 1** The virus that causes oral herpes, which is characterized by cold sores or fever blisters on the lips or mouth. Abbreviated *HSV-1*.

**Herpes simplex virus type 2** The virus that causes genital herpes. Abbreviated *HSV-2*.

**Human immunodeficiency virus (HIV)** A sexually transmitted virus that destroys white blood cells in the immune system, leaving the body vulnerable to life-threatening diseases.

**Human papilloma virus** The virus that causes genital warts. Abbreviated *HPV*.

**Immune system** A term for the body's complex of mechanisms for protecting itself from disease-causing agents such as pathogens.

**Injectable drug users** Persons who inject drugs into their veins (intravenous drug users) or under the skin. Abbreviated *IDUs*.

**Leukocytes** White blood cells which are essential to the body's defenses against infection. (From the Greek *leukos*, meaning "white" and *kytos*, meaning "a hollow" and used in combination with other word forms to mean *cell*.)

**Lymphogranuloma venereum** A tropical STD caused by the *Chlamydia trachomatis* bacterium.

**Molluscum contagiosum** An STD is caused by a pox virus that causes painless raised lesions to appear on the genitals, buttocks, thighs, or lower abdomen.

**Ocular herpes** A herpes infection of the eye, usually caused by touching an infected area of the body and then touching the eye.

**Opportunistic diseases** Diseases that take hold only when the immune system is weakened and unable to fend them off. Kaposi's sarcoma and pneumocystis carinii pneumonia (PCP) are examples of opportunistic diseases found in AIDS patients.

**Outercourse** Forms of sexual expression, such as massage, hugging, caressing, mutual masturbation, and rubbing bodies together that do not involve the exchange of body fluids. (Contrast with *intercourse*.)

**Pathogens** Agents, especially microorganisms, that can cause a disease. (From the Greek *pathos*, meaning "suffering" or "disease," and *genic*, meaning "forming" or "coming into being.")

**Pediculosis** A parasitic infestation by pubic lice (*Pthirus pubis*) that causes itching.

**Pelvic inflammatory disease** Inflammation of the pelvic region-- possibly including the cervix, uterus, fallopian tubes, abdominal cavity, and ovaries--that can be caused by the gonoccocal bacterium or other organisms. The condition may lead to infertility and cause such symptoms as abdominal pain, tenderness, nausea, fever, and irregular menstrual cycles. Abbreviated *PID*.

**Prophylactic** A drug or device that provides protection from disease.

**Pubic lice** Insects that belong to a family of biting lice. (Singular: pubic louse)

**Scabies** A parasitic infestation caused by a tiny mite (*Sarcoptes scabiei*) that causes itching.

**Sexually transmitted diseases** Diseases transmitted through sexual means (formerly called venereal diseases or VD). Abbreviated *STDs*.

**Shigellosis** An STD caused by the *Shigella* bacterium.

**Syphilis** An STD that is caused by the *Treponema pallidum* bacterium and that may progress through several stages of development--often from a chancre to a skin rash to damage to the cardiovascular or central nervous systems. (From the Greek *siphlos*, meaning "maimed" or "crippled.")

**Trichomoniasis** A form of vaginitis caused by the protozoan *Trichomonas vaginalis*.

**Vaginitis** Any type of vaginal infection or inflammation.

**VDRL** The test named after the Venereal Disease Research Laboratory of the U.S. Public Health Service that tests for the presence of antibodies in the blood to *Treponema pallidum*.

**Viral hepatitis** Hepatitis caused by one of several types of viruses.

# NOTES

## Foreword

1.     Gayle, H. D., et al. (1990). Prevalence of human immunodeficiency virus among university students. *The New England Journal of Medicine, 323*, 1538-1541.

## Chapter 1

1.     From *TELLING IT LIKE IT IS: STRAIGHT TALK ABOUT SEX* by Marjorie Brant Osterhout, Annamaria Formichella, and Susan McIntyre, developed by Dr. Sean Gresh. Copyright ©1991 by Majorie Brant Osterhout, et al. Published by arrangement with AVON BOOKS, a division of the Hearst Corporation. (Preprint ed., pp. 137-139.)
2.     Shelton, D. "STDs: A Hidden Epidemic." *American Medical News,* American Medical Association, December 9, 1996.
3.     Shelton (1996)
4.     Johnson, D. (1990, March 8). AIDS clamor at colleges muffling older dangers. *The New York Times*, p. A18.
5.     American Social Health Association. (1991). *STD (VD)*. Research Triangle Park, NC: Author.
6.     Yarber, W. L. (1985). *STD: A guide for today's young adults.* Waldorf, MD: American Alliance Publications.
7.     Centers for Disease Control [CDC], Division of STD/HIV Prevention. (1990). *Annual report.*
8.     Rathus, S. A., Nevid, J.S., & Fichner-Rathus, L. (1997). *Human sexuality in a world of diversity.* (3rd Ed.). Boston: Allyn & Bacon.
9.     Reinisch, J. M. (1990). *The Kinsey Institute new report on sex: What you must know to be sexually literate.* New York: St. Martin's Press.
10.    Drinnin, B. (1993). *Instructors annotated edition, Human sexuality in a world of diversity.* Boston: Allyn & Bacon.
11.    Drinnin (1993)
12.    Goldsmith, M. F. (1989). Medical news and perspectives: "Silent epidemic" of "social disease" makes STD experts raise their voices. *Journal of the American Medical Association, 261*, 3509-3510.
13.    Boston Women's Health Book Collective. (1984). *The New Our Bodies, Ourselves.* New York: Simon & Schuster.
14.    Johnson (1990)
15.    Rathus, S. A., & Nevid, J. S. (1992). *Adjustment and growth: The challenges of life* (5th ed.). Fort Worth, TX: Harcourt Brace Jovanovich. p. 487. Reprinted with permission.
16.    American Social Health Association, 1991. Reprinted with permission.
17.    MacDonald, N. E., et al. (1990). High-risk STD/HIV behavior among college students. *Journal of the American Medical Association, 263*, 3155-3159.

## Chapter 2

1.     Gottlieb, M. S. (1991, June 5). AIDS—the second decade: Leadership is lacking. *The New York Times*, p. A29.
2.     Shilts, R. (1987). *And the band played on: Politics, people, and the AIDS epidemic.* New York: Penguin Books.

3.  Jones, D. S., et al. (1992). Epidemiology of transfusion-associated acquired immunodeficiency syndrome in children in the United States, 1981 through 1989. *Pediatrics, 89,* 123-127.

4.  Altman, L. K. Experiment to See if AIDS Can Be Cured Is Delayed a Year. *The New York Times,* January 23, 1997.

5.  Chartrand, S. (1996, August 12). Mixers for 'Cocktails' Used to Delay AIDS. *The New York Times,* p. D2.

6.  Essex, M., & Kanki, P. (1988, October). The origins of the AIDS virus. *Scientific American,* 64-71; Norman, C. (1986). Politics and science clash on African AIDS. *Science, 230,* 1140-1141; Smith, T. F., et al. (1988). The phylogenetic history of immunodeficiency viruses. *Nature, 33,* 573-575.

7.  Kolata, G. (1991b, November 28). Theory links AIDS to malaria experiments. *The New York Times,* p. B14.

8.  Cowley, G. (1993, March 22). The future of AIDS. *Newsweek,* pp. 46-52.

9.  Kolata, G. (1991c, June 4). 10 years of AIDS battle: Hopes for success dim. *The New York Times,* p. A14; Altman (1991)

10. Gottlieb (1991)

11. Kramer, L. (1990, July 16). A "Manhattan Project" for AIDS. *The New York Times,* p. A15.

12. Reported by the Associated Press, February 28, 1997; Altman, L. K. (1993a, June 6). At AIDS talks, science confronts daunting maze. *The New York Times,* p. A20.

13. Hammond, T. The Estimated Prevalence of HIV in the United States: Signs of Decline. *American Journal of Public Health,* May 1996.

14. World Health Organization. (1995, September 12). Cited in "Rise in STDs concerns group." *Newsday,* p. B27

15. Altman (1991)

16. Mann, J., Tarantola, D., & Netter, T.W. (1992). *AIDS in the world: A global report.* Cambridge, MA: Harvard University Press.

17. Lewis, P. (1992, April 16). Japanese live longer, the U.N. finds. *The New York Times,* p. A12.

18. Altman, L. K. (1993d, July 23). Sex is leading cause of AIDS in women. *The New York Times,* p. A12; CDC (1993a). Update: Acquired immunodeficiency syndrome--United States, 1992. *Mortality and Morbidity Weekly Report, 42 (No. 28),* pp. 547-557.

19. CDC (1996). *HIV/AIDS Surveillance Report: U.S. HIV and AIDS cases reported through June 1996,* Vol.8, No.1.

20. CDC (1990b). Heterosexual behaviors that influence condom use among patients attending a sexually transmitted disease clinic—San Francisco. *Morbidity and Mortality Weekly Report, 39,* 685-689; Cowley (1993); Ehrhardt, A. A. (1992). Trends in sexual behavior and the HIV pandemic. *American Journal of Public Health, 82,* 1459-1461. (Editorial)

21. Ehrhardt (1992)

22. DesJarlais, D. C., & Friedman, S. R. (1988). The psychology of preventing AIDS among intravenous drug users: A social learning conceptualization. *American Psychologist, 43,* 865-871.

23. Specter, M. (1991, November 8). Magic's loud message for young black men. *The New York Times,* p. B12.

24. Goodgame, R. W. (1990). AIDS in Uganda—Clinical and social features. *The New England Journal of Medicine, 323,* 383-389; Quinn, T. C. (1990). Unique aspects of human immunodeficiency virus and related viruses in developing countries. In K. K. Holmes, et al. (Eds.), *Sexually transmitted diseases* (2nd ed.) (pp. 355-369). New York: McGraw-Hill, Inc.

25. Novello, A. C. (1991). Women and HIV infection. *Journal of the American Medical Association, 265,* 1805.

26. Haverkos, H. W. (1993). Reported cases of AIDS: An update. *The New England Journal of Medicine, 329,* 511.

27. CDC (1990b)

28. MacDonald et al. (1990)

29.  Carroll, L. (1988). Concern with AIDS and the sexual behavior of college students. *Journal of Marriage and the Family, 50,* 405-411.

30.  DeBuono, B. A., et al. (1990). Sexual behavior of college women in 1975, 1986, and 1989. *The New England Journal of Medicine, 322,* 821-825.

31.  Johnson (1990)

32.  Hernandez, J. T., & Smith, F. J. (1990). Inconsistencies and misperceptions putting college students at risk of HIV infection. *Journal of Adolescent Health Care, 11,* 295-297.

33.  Oswalt, R., & Matsen, K. (1993). Sex, AIDS, and the use of condoms : A survey of compliance in college students. *Psychological Reports, 72,* 764-766.

34.  Adams, D. D., et al. (1991, Fall). Southern college students: Beliefs about AIDS. *Journal of Sex Education and Therapy, 17.*

35.  Nyamathi, A., et al. (1995). Psychosocial predictors of AIDS risk behavior and drug use behavior in homeless and drug addicted women of color. *Health Psychology, 14,* 265-273.

36.  St. Lawrence, J. S., et al. (1995). Cognitive-behavioral intervention to reduce African American adolescents' risk for HIV infection. *Journal of Consulting and Clinical Psychology, 63,* 221-237.

37.  Cochran, S. D., & Mays, V. M. (1989). Women and AIDS-related concerns: Roles for psychologists in helping the worried well. *American Psychologist, 44,* 529-535.

## Chapter 3

1.  Vaccination is the placement of a weakened form of an antigen in the body, which activates the creation of antibodies and memory lymphocytes. Smallpox has been annihilated by vaccination, and researchers are trying to develop a vaccine against the virus that causes AIDS.

2.  Shilts (1987)

3.  Reinisch (1990)

4.  Navarro, M. (1992, February 10). Agencies hindered in effort to widen definitions of AIDS. *The New York Times,* pp. 1, B11.

5.  Hamilton, J. D., et al. (1992). A controlled trial of early versus late treatment with zidovudine in symptomatic human immunodeficiency virus infection. *The New England Journal of Medicine, 326,* 437-443.

6.  Hatcher, R. A., et al. (1990). *Contraceptive technology: 1990—1992* (15th rev. ed.). New York: Irvington Publishers.

7.  CDC (1992). *HIV infection and AIDS: Are you at risk?* Atlanta: Author.

8.  Balter, M. AIDS Research: New Hope in HIV Disease. *Science, 275* (1997).

9.  Glasner, P. D., & Kaslow, R. A. (1990). The epidemiology of human immunodeficiency virus infection. *Journal of Consulting and Clinical Psychology, 58,* 13-21; Jones et al., 1992; Simonds et al. (1992). Transmission of human immunodeficiency virus type 1 from a seronegative organ and tissue donor. *The New England Journal of Medicine, 326,* 726-732.

10.  CDC (1992)

11.  Padian, N. S., Shiboski, S. C., & Jewell, N. P. (1991). Female-to-male transmission of human immunodeficiency virus. *Journal of the American Medical Association, 266,* 1664-1667.

12.  Allen, J. R., & Setlow, V. P. (1991). Heterosexual transmission of HIV: A view of the future. *Journal of the American Medical Association, 266,* 1695-1696; Rosenthal, E. (1990, August 28). The spread of AIDS: A mystery unravels. *The New York Times,* pp. C1, C2

13.  Allen & Setlow (1991)

14.  Perry, S., Jacobsberg, L., & Fogel, K. (1989). Orogenital transmission of human immunodeficiency virus (HIV). *Annals of Internal Medicine, 111,* 951-952; Spitzer, P. G., & Weiner, N. J. (1989). Transmission of HIV infection from a woman to a man by oral sex. *Journal of the*

*American Medical Association, 320,* 251.

15. CDC (1992b, January). *HIV infection and AIDS: Are you at risk?* Atlanta: Author.
16. Rosenthal (1990)
17. Quinn (1990)
18. Rosenthal (1990)
19. Holmes, K. K., & Kreiss, J. (1988). Heterosexual transmission of human immunodeficiency virus: Overview of a neglected aspect of the AIDS epidemic. *Journal of Acquired Immune Deficiency Syndromes, 1,* 602-610; Simonsen, J. N., et al. (1988). Human immunodeficiency virus infection among men with sexually transmitted diseases: Experience from a center in Africa. *The New England Journal of Medicine, 319,* 274-278.
20. Touchette, N. (1991). HIV-1 link prompts circumspection of circumcision. *The Journal of NIH Research, 3,* 44-46.
21. CDC (1992)
22. DesJarlais & Friedman (1988); Drug and Sex Programs Called Effective in Fight Against AIDS. (1997, February 14). *The New York Times,* p. A27.
23. Friedman, S. R., et al. (1987). AIDS and self-organization among intravenous drug uses. *International Journal of Addictions, 22,* 201-220.
24. Peckham, C., & Gibb, D. (1995). Mother-to-child transmission of the human immunodeficiency virus. *The New England Journal of Medicine, 333,* 298-302.
25. CDC (1990a); Gwinn, M., et al. (1991). Prevalence of HIV infection in childbearing women in the United States: Surveillance using newborn blood samples. *Journal of the American Medical Association, 265,* 1704-1708.
26. CDC, Division of Sexually Transmitted Diseases/HIV Prevention (1991). *Annual Report.* Atlanta: Author.
27. Goedert, J. J., et al. (1992, January 7). Cited in Leary, W. E. Study of H.I.V. transmission at birth. *The New York Times,* p. C3.
28. CDC (1992)
29. Connor, E. M., et al. (1994). Reduction of maternal-infant transmissin of human immunodeficiency virus type I with zidovudine treatment. The New England Journal of Medicine, *331,* 1173-1180.
30. Fogle, S. (1991). AIDS hemophiliacs in tough court battles. *The Journal of NIH Research, 3,* 46-47.
31. Fogle (1991)
32. Altman, L. K. (1992, April 9). Ashe received a transfusion before blood supply was tested for H.I.V. *The New York Times,* B15.
33. Altman (1992)
34. Adler, J., et al. (1991, November 18). Living with the virus: When—and—how—HIV turns into AIDS. *Newsweek,* pp. 33-34; Lambert, B. (1991, December 9). Kimberly Bergalis is dead at 23; symbol of debate over AIDS tests. *The New York Times,* p. D9.
35. Altman, L. K. (1992b, May 15). Study sees no new transmission of HIV by health-care workers. *The New York Times,* p. A18.
36. Taylor, R. Dentist-to-patient HIV-1 transmission.: More heat, No light. *Journal of NIH Research,* July 1993, p. 50.
37. Simons, M. (1997, January 22). French medical group asks doctors with H.I.V. to halt surgery. *The New York Times,* p. A5.
38. Lambert (1991)
39. Gordon & Snyder (1989) *Personal issues in human sexuality: A guidebook for better sexual health* (2nd ed.). Boston: Allyn & Bacon; Hatcher, R. A., et al. (1990)
40. Hatcher et al. (1990)
41. CDC (1992)
42. Cochran & Mays, 1989; CDC (1992)

43. CDC (1992); Hatcher et al. (1990); Holmberg & Curran (1989). The epidemiology of HIV infection in industrialized countries. In K.K. Holmes, et al. (Eds.), *Sexually transmitted diseases* (pp. 343-354). New York: McGraw-Hill, Inc.
44. Holmberg & Curran (1989)
45. AIDS without needles or sex. (1993, December 20). *Newsweek,* pp. 106-107.
46. CDC (1988b). Leads from the MMWR/Morbidity and Mortality Weekly Report (Vol 37/717-727): HIV-related beliefs, knowledge, and behaviors among high school students. *Journal of the American Medical Association, 260,* 3567, 3570.
47. Sonenstein, F. L., Pleck, J. H., & Ku, L. C. (1989). Sexual activity, condom use and AIDS awareness among adolescent males. *Family Planning Perspectives, 21,* 152-157.
48. CDC (1988b)
49. DeBuono et al. (1990); O'Gorman, E. C., & Bownes, I. T. (1990). Factors influencing behavioural change in response to AIDS educational programmes—the role of cognitive distortions. *Medical Science Research, 18,* 263-264; Ruder, A. M., et al. (1990). AIDS education: Evaluation of school and worksite based presentations. *New York State Journal of Medicine, 90,* 129-133.
50. Kegeles, S. M., Alan, M. E., & Irwin, C. (1988). Sexually active adolescents and condoms: Changes over one year in knowledge, attitudes, and use. *American Journal of Public Health, 78,* 460-461; Leishman, K. (1987, February). Heterosexuals and AIDS. *The Atlantic Monthly,* pp. 39-58.
51. CDC (1992)
52. CDC (1992); CDC. *HIV infection and AIDS: Are you at risk?* NAIEP, September 1994, # D539.

**Chapter 4**

1. Adapted from Rathus & Nevid, 1992, pp.485-486.
2. Nyamathi et al. (1995)
3. Reported by the Associated Press, February 28, 1997
4. CDC (1991); Kent, M. R. (1991). Women and AIDS. *New England Journal of Medicine, 324,* 1442.
5. Altman (1992b)
6. Ellerbrock, T. V., et al. (1991). Epidemiology of women with AIDS in the United States, 1981 through 1990: A comparison with heterosexual men with AIDS. *Journal of the American Medical Association, 265,* 2971-2975.
7. Parsons, E. (1991, November 18). Women become top U.S. AIDS risk group. *The New York Times,* p. A14. Copyright © 1991 by the New York Times Company. Reprinted by permission.
8. Amaro, H. (1995). Love, sex, and power: Considering women's realities in HIV prevention. *American Psychologist, 50,*437-447.
9. CDC (1991); Ellerbrock et al. (1991); Toll of American AIDS orphans put at 80,000 by end of decade. (1992, December 23). *The New York Times,* p. B6.
10. Chu, S.Y., Buehler, J.W., & Berkelman, R.L. (1990). Impact of human immunodeficiency virus epidemic on mortality in women of reproductive age, United States. *Journal of theAmerican Medical Association, 264,* 225-229.
11. CDC (1991)
12. CDC (1991)
13. Altman (1993); CDC (1993a)
14. Kent ( 1991); Stephens, T. (1991). AIDS in women reveals health-care deficiencies. *The Journal of NIH Research, 3,* 27-30.
15. Cohen, J. B. (1990, December 13). *A crosscutting perspective on the epidemiology of HIV infection in women.* Paper Presented at the Women and AIDS Conference, Washington, DC (abstract).

16. Kent (1991)
17. Stephens (1991)
18. Amaro (1995); Weinstock, H.S., et al. (1993). Factors associated with condom use in a high-risk heterosexual population. *Sexually Transmitted Diseases, 20,* 14-20.

**Chapter 5**

1. Osterhout et al. (1991), pp. 145-147.
2. Kubic, M. New ways to prevent and treat AIDS. *FDA Consumer*, January-February 1997.
3. CDC (1990f). Update: Serotologic testing for HIV-1 antibody—United States, 1988-1989. *Morbidity and Mortality Weekly Report, 39,* 380-383.
4. Reinisch (1990)
5. Burke, D. S., et al. (1988). Measurement of the false positive rate in a screening program for human immunodeficiency virus infections. *New England Journal of Medicine, 319,* 961-964.
6. Some of the questions here are modeled after Osterhout et al., 1991.
7. Altman, L. K. (1993a)
8. Altman, L. K. (1993a)
9. Cooper, D. A., et al. (1993). Zidovudine in persons with asymptomatic HIV infection and CD4 cell counts greater than 400 per cubic millimeter." *New England Journal of Medicine, 329,* 297-303; Kinloch-de Loes, S., et al. (1995). A controlled trial of zidovudine in primary human immunodeficiency virus infection. *New England Journal of Medicine, 333,* 408-413.
10. Yarchoan, R., Mitsuya, H., & Broder, S. (1988). AIDS therapies. *Scientific American, 259,* 110-119.
11. Hammer, S. M.., et al. (1996). A trial comparing nucleoside monotherapy with combination therapy in HIV-infected adults with CD4 cell counts from 200 to 500 per Cubic Millimeter." *New England Journal of Medicine, 335,* 1081-1090; Katzenstein, D. A., et al. (1996). The relation of virologic and immunologic markers to clinical outcomes after nucleloside therapy in HIV-infected adults with 200 to 500 CD4 cells per cubic millimeter. *New England Journal of Medicine, 335,* 1091-1098.
12. Corey, L., & Holmes, K. K. (1996). Therapy for HIV Infection--What have we learned?" *New England Journal of Medicine, 335,* 1142-1144; Deeks, S. G., et al. (1997). HIV-1 protease inhibitors: A Review for Clinicians. *Journal of the American Medical Association, 277,* 145-153.
13. Balter, M. (1997). AIDS Research: New Hope in HIV Disease." *Science, 275,* in press.
14. Chartrand, S. (1996). Mixers for cocktails used to delay AIDS. *The New York Times*, August 12, 1996, p.D2.
15. Altman, L. K. (1997, January 23). Experiment to see if AIDS can be cured is delayed a year. *The New York Times*, p C3.
16. Reported by the Associated Press, February 28, 1997
17. Altman, L. K. (1997, January 25). Deaths from AIDS decline sharply in New York City. *The New York Times*, pp. A1, A28.
18. Reported by the Associated Press, February 28, 1997.
19. Kolata, G. (1996, September 15). AIDS patients slipping through safety net. *The New York Times*, p. A24; Balter, 1997; Pear, R. (1997, February 16). Expense means many can't get drugs for AIDS. *The New York Times*, pp. A1, A36.

**Chapter 6**

1. ABC List. American Social Health Association, 1996.
2. National Center for Health Statistics. (1985). *Health: United States 1985*. Washington, D.C.:

U.S. Department of Health and Human Services.

3.  Calderone, M. S., & Johnson, E. W. (1989). *Family book about sexuality* (rev. ed.). New York: Harper & Row; Reinisch (1990)

4.  Calderone & Johnson (1989)

5.  Reinisch (1990)

6.  Handsfield, H. (1984). Gonorrhea and uncomplicated gonococcal infection. In K. K. Holmes, et al. (Eds.), *Sexually transmitted diseases* (pp. 205-220). New York: McGraw-Hill; Platt, R., Rice, P., & McCormack, W. (1983). Risk of acquiring gonorrhea and prevalence of abnormal adrenal findings among women recently exposed to gonorrhea. *Journal of the American Medical Association, 250,* 3205-3209.

7.  Reinisch (1990)

8.  National Institutes of Health, National Institute of Allergy and Infectious Diseases. *Pelvic Inflammatory Disease.* Office of Communications National Institute of Allergy and Infectious Diseases National Institutes of Health, August 1992.

9.  Reinisch (1990)

10. Ison, C. A. (1990). Laboratory methods in genitourinary medicine: Methods of diagnosing gonorrhoea. *Genitourinary Medicine, 66,* 453-459; Judson, F. N. (1990). Gonorrhea. *Medical Clinics of North America, 74,* 1353-1366.

11. Goldstein, A. M., & Clark, J. H. (1990). Treatment of uncomplicated gonococcal urethritis with single-dose ceftriaxone. *Sexually Transmitted Diseases, 17,* 181-183.

12. Goldstein & Clark (1990)

13. CDC (1989b). Treatment guidelines for sexually transmitted diseases. *Morbidity and Mortality Weekly Report, 38,* No. S-8.

14. Judson (1990)

15. Moran, J. S., & Zenilman, J. M. (1990). Therapy for gonococcal infections: Options in 1989. *Reviews of Infectious Diseases,* (Suppl. 6.) S633-S644.

## Chapter 7

1.  Zenker, P. N., & Rolfs, R. T. (1990). Treatment of syphilis, 1989. *Reviews of Infectious Diseases* (Suppl. 6), S590-S609.

2.  Centers for Disease Control and Prevention. Sexually Transmitted Disease Surveillance 1995, Division of STD Prevention, September 1996.

3.  American Social Health Association (1996)

4.  Minkoff et al. (1990); Rolfs, R. T., Goldberg, M., & Sharrar, R. G. (1990) Risk factors for syphilis: Cocaine use and prostitution. *American Journal of Public Health, 80,* 853-857.

5.  Rolfs et al. (1990)

6.  Farley, A. U., Hadler, J. L., & Gunn, R. A. (1990). The syphilis epidemic in Connecticut: Relationship to drug use and prostitution. *Sexually Transmitted Diseases, 17,* 163-168.

7.  Reinisch (1990)

8.  Wooldridge, W. E. (1991). Syphilis: A new visit from an old enemy. *Postgraduate Medicine, 89,* 199-202.

9.  CDC (1989b)

## Chapter 8

1. USDHHS (1992). *What we have learned from the AIDS community Demonstration Projects.* Atlanta: Center for Disease Control.
2. ABC List. American Social Health Association, 1996.
3. Shafer, M. A., et al. (1993). Evaluation of urine-based screening strategies to detect *Chlamydia trachomatis* among sexually active asymptomatic young men. *Journal of the American Medical Association, 270,* 2065-2070. CDC (1993b). Evaluation of surveillance for *Chlamydia trachomatis* infections in the United States, 1987 to 1991. *Mortality and Morbidity Weekly Report, 42 (SS-3),* 21-27.
4. Yarber, W. L., & Parillo, A. V. (1992). Adolescents and sexually transmitted diseases. *Journal of School Health, 62,* 331-338.
5. Martin, D. H. (1990). Chlamydial infections. *Medical Clinics of North America, 74,* 1367-1387.
6. Martin (1990); Westrom, L. V. (1990). Chlamydia trachomatis—clinical significance and strategies of intervention. *Seminars in Dermatology, 9,* 117-125.
7. Braude, A. I., Davis, C. E., & Fierer, J. (Eds.). (1986). *Infectious diseases and medical microbiology* (2nd ed.) Philadelphia: W. B. Saunders.
8. CDC (1989b)
9. CDC (1985b). Chlamydia Trachomatis infection. *Morbidity & Mortality Weekly Report, 34,* 53.
10. Bowie, W. R. (1990). Approach to men with urethritis and urologic complications of sexually-transmitted diseases. *Medical Clinics of North America, 74,* 1543-1557.
11. Bowie (1990)
12. CDC (1989b)
13. Stamm, W. E., & Holmes, K. K. (1990). *Chlamydia trachomatis* infections of the adult. In K. K. Holmes et al. (Eds.), *Sexually transmitted diseases.* (2nd ed.) (pp. 181-194). New York: McGraw-Hill Information Services Company.
14. American Social Health Association (1996)
15. Handsfield, H. H. (1988). Questions and answers: "Safe sex" guidelines: Mycoplasma and chlamydia infections. *Journal of the American Medical Association, 259,* 2022.
16. Bowie, W. (1984). Epidemiology and therapy of chlamydia trachomatis infections. *Drugs, 27,* 459-468.
17. Schachter, J. (1990). Biology of *Chlamydia trachomatis.* In K. K. Holmes, P. Mardh, P. F. Sparling, & P. J. Wiesner (Eds.). *Sexually transmitted diseases* (2nd Ed.). (pp. 161-180). New York: McGraw-Hill.
18. Reinisch (1990)
19. Graham, J. M., & Blanco, J. D. (1990). Chlamydial infections. *Primary Care: Clinics in Office Practice, 17,* 85-93.
20. Reinisch (1990)
21. CDC. Sexually Transmitted Disease Surveillance 1995, September 1995; Hodgson, R., et al. (1990). Chlamydia trachomatis: The prevalence, trend and importance in initial infertility management. *Australian and New Zealand Journal of Obstetrics and Gynaecology, 30,* 25-254; Garland, S. M., Lees, M. I., & Skurrie, I. J. (1990). Chlamydia trachomatis: Role in tubal infertility. *Australian and New Zealand Journal of Obstetrics and Gynecology, 30,* 83-86.
22. American Social Health Association, 1996
23. Sherman, K. J., et al. (1990). Sexually transmitted diseases and tubal pregnancy. *Sexually Transmitted Diseases, 17,* 115-121.
24. Crum, C., & Ellner, P. (1985). Chlamydia infections: Making the diagnosis. *Contemporary Obstetrics and Gynecology, 25,* 153-159, 163, 165, 168.
25. Bowie (1990)
26. Reichart, C. A., et al. (1990). Evalution of Abbott Testpack Chlamydia for detection of chlamydia

trachomatis in patients attending sexually transmitted diseases clinics. *Sexually Transmitted Diseases, 17*, 147-151.

27. Centers for Disease Control and Prevention. 1993 Sexually transmitted diseases treatment guidelines. *MMWR* , 1993, 42(No. RR-14).
28. Martin (1990); CDC (1993a)
29. Stamm & Holmes (1990)
30. CDC (1993a)

### Chapter 9

1. Sobel, J. D. (1990). Vaginal infections in adult women. *Medical Clinics of North America, 74*, 1573-1602.
2. Friedrich, E. (1985). Vaginitis. *American Journal of Obstetrics and Gynecology, 152*, 247-251.
3. Reinisch (1990)
4. Briselden, A. M., & Hillier, S. L. (1990). Longitudinal study of the biotypes of Gardnerella vaginalis. *Journal of Clinical Microbiology, 28*, 2761-2764; Platz-Christensen, J., et al. (1989). Detection of bacterial vaginosis in Papanicolaou smears. *American Journal of Obstetrics and Gynecology, 160*, 132-133.
5. Hillier, S., & Holmes, K. K. (1990). Bacterial vaginosis. In K. K. Holmes, P. Mardh, P. F. Sparling, & P. J. Wiesner (Eds.). *Sexually transmitted diseases* (2nd Ed.). (pp. 547-560). New York: McGraw-Hill.
6. Hillier & Holmes (1990)
7. Hillier & Holmes (1990)
8. Reinisch (1990)
9. Sobel (1990)
10. Vaginal yeast infection can be an HIV warning. (1992, November 24). *New York Newsday*, p. 51.
11. Sobel (1990)
12. Friedrich (1985)
13. Hilton, E., et al. (1992). Ingestion of yogurt containing Lactobacillus acidophilus as prophylaxis for candidal vaginitis. *Annals of Internal Medicine, 116*, 353-357.
14. Sobel (1990)
15. Levy, M. R., Dignan, M., & Shirreffs, J. H. (1987). *Life and health* (5th ed.). New York: Random House.
16. Sobel (1990)
17. Levy et al. (1987)
18. Levine, G. I. (1991). Sexually transmitted parasitic diseases. *Primary Care: Clinics in Office Practice, 18*, 101-128.
19. Martens, M., & Faro, S. (1989, January). Update on trichomoniasis: Detection and management. *Medical Aspects of Human Sexuality*, 73-79.
20. Rein, M. F., & Müller, M. (1990). *Trichomonas vaginalis* and trichomoniasis. In K. K. Holmes, P. Mardh, P. F. Sparling, & P. J. Wiesner (Eds.). *Sexually transmitted diseases* (2nd Ed.) (pp. 481-492). New York: McGraw-Hill.
21. Rein & Müller (1990)
22. Rein & Müller (1990)
23. Grodstein, F., Goldman, M. G., & Cramer, D. W. (1993). Relation of tubal infertility to history of sexually transmitted diseases. *American Journal of Epidemiology, 137*, 577-584.
24. Levine (1991)
25. Thomason, J. L., & Gelbart, S. M. (1989). Trichomonas vaginalis. *Obstetrics and Gynecology, 74*, 536-541.

26. Rein & Müller (1990)
27. Reinisch (1990)
28. Reinisch (1990)
29. Rein & Müller (1990)
30. Reinisch (1990)
31. CDC (1989b)
32. CDC (1989b)
33. Reinisch (1990)
34. CDC (1989b); Moi, H., et al. (1989). Should male consorts of women with bacterial vaginosis be treated? *Genitourinary Medicine, 65,* 263-268; Hillier & Holmes (1990)
35. CDC (1989b)
36. Sobel (1990)
37. Reinisch (1990); Thomason & Gelbart (1989)
38. Boston Women's Health Book Collective (1984), p. 518; Adapted from Rathus et al. (1993)

## Chapter 10

1.  Braude et al. (1986); Kunz, J. R. M., & Finkel, A. J. (1987) (Eds.) *The American Medical Association family medical guide: Revised and updated.* New York: Random House.
2.  American Social Health Association (1996)
3.  American Social Health Association (1996)
4.  Straus, S. E. (1985). Herpes simplex virus infections: Biology, treatment, and prevention. *Annals of Internal Medicine, 103,* 404-419.
5.  Corey, L. (1990). Genital herpes. In K. K. Holmes, et al. (Eds.), *Sexually transmitted diseases* (2nd ed.) (pp. 391-414). New York: McGraw-Hill, Inc.
6.  Brody, J.E. (1992b, August 12). Genital herpes thrives on ignorance and secrecy. *The New York Times*, p. C12; Fox, M. (1985, December). Interfering with herpes. *Today's Health*, 22.
7.  Peterman, T., Cates, W., & Curran, J. (1988). The challenge of human immunodeficiency virus (HIV) and acquired immunodeficiency syndrome (AIDS) in women and children. *Fertility and Sterility, 49,* 571-581.
8.  Brock, B. V., et al. (1990). Frequency of asymptomatic shedding of herpes simplex virus in women with genital herpes. *The Journal of the American Medical Association, 263,* 418-420; Dawkins, B. J. (1990). Genital herpes simplex infections. *Primary Care: Clinics in Office Practice, 17,* 95-113; Rooney, J., et al. (1986). Acquisition of genital herpes from an asymptomatic sexual partner. *New England Journal of Medicine, 314,* 1561-1564.
9.  Brock et al. (1990)
10. Whitley, R., et al. (1991). Predictors of morbidity and mortality in infants with herpes simplex virus infections. *The New England Journal of Medicine, 324,* 450-454.
11. Dawkins (1990); Osborne, N. G., & Adelson, M. D. (1990). Herpes simplex and human papillomavirus genital infections: Controversy over obstetric management. *Clinical Obstetrics and Gynecology, 33,* 801-811.
12. Brown, Z. A., et al. (1991). Neonatal herpes simplex virus infection in relation to asymptomatic maternal infection at the time of labor. *The New England Journal of Medicine, 324,* 1247-1252.
13. Campbell, C. E., & Herten, R. J. (1981). VD to STD: Redefining venereal disease. *American Journal of Nursing, 81,* 1629-1635.
14. Graham, S. et al. (1982). Sex patterns and herpes simplex virus type 2 in the epidemiology of cancer of the cervix. *American Journal of Epidemiology, 115,* 729-735; Hatcher et al. (1990)
15. Straus, S. E., et al. (1984a). Suppression of frequently recurring genital herpes: A placebo-

controlled double-blind trial of oral acyclovir. *New England Journal of Medicine, 310,* 1545-1550.

16. Brody (1992b)
17. Kemeny, M. E., Cohen, F., Zegans, L. S., & Conant, M. A. (1989). Psychological and immunological predictors of genital herpes recurrence. *Psychosomatic Medicine, 51,* 195-208; Kunz & Finkel (1987); Longo, D. J., & Clum, G. A. (1989). Psychosocial factors affecting genital herpes recurrences: Linear vs. mediating models. *Journal of Psychosomatic Research, 33,* 161-166.
18. Rand, K. H., et al. (1990). Daily stress and recurrence of genital herpes simplex. *Archives of Internal Medicine, 150,* 1889-1893.
19. Corey (1990)
20. Reinisch (1990)
21. Reinisch (1990)
22. Mertz, G. J. (1990). Genital herpes simplex virus infections. *Medical Clinics of North America, 74,* 1433-1454.
23. Gold, D., et al. (1988). Chronic-dose acyclovir to suppress frequently recurring genital herpes simplex virus infection: Effect on antibody response to herpes simplex virus type 2 proteins. *Journal of Infectious Diseases, 158,* 1227-1234; Goldberg, L. H., et al. (1993). Longterm suppression of recurrent genital herpes with acyclovir. *Archives of Dermatology, 129,* 582-587.
24. Brody (1993e); Goldberg et al. (1993)
25. Kaplowitz, L. G., et al. (1991). Prolonged continuous acyclovir treatment of normal adults with frequently recurring genital herpes simplex virus infection. *Journal of the American Medical Association, 265,* 747-751.
26. Hatcher et al. (1990); Stone, K. M., & Whittington, W. L. (1990). Treatment of genital herpes. *Reviews of Infectious Diseases* (Suppl. 6), S633-S644.
27. Levy et al. (1987); Mirotznik, J., et al. (1987). Genital herpes: An investigation of its attitudinal and behavioral correlates. *Journal of Sex Research, 23,* 266-272.
28. Mirotznik et al. (1987)
29. Boston Women's Health Book Collective (1984), *The New Our Bodies, Ourselves,* p. 277
30. Laskin, D. (1982, February 21). The herpes syndrome. *The New York Times Sunday Magazine,* pp. 94-108.

## Chapter 11

1. NIAID. *Hepatitis.* August, 1992.
2. Kunz & Finkel (1987)
3. Kunz & Finkel (1987)
4. Brody, J. E. (1997, January 22). There is bad news and good about a hidden viral epidemic: Hepatitis C. *The New York Times,* p. C9; Experts warn of a tripling of deaths from hepatitis C by 2017. (1997, March 27). *The New York Times,* p. A21.
5. NIAID (1992)
6. Lemon, S. J., & Newbold, J. E. (1990). Viral hepatitis. In K. K. Holmes, et al. (Eds.), *Sexually transmitted diseases* (2nd ed.) (pp. 449-466). New York: McGraw-Hill, Inc.
7. Lemon & Newbold (1990)
8. Lemon & Newbold (1990)
9. CDC (1989b)
10. Lemon & Newbold (1990)
11. Alter, M. J., et al.. (1989). The importance of heterosexual activity and intravenous drug use in the transmission of hepatitis B and non-A, non-B hepatitis. *Journal of the American Medical Association, 262,* 1201-1205.

12. Davis, G. L., et al. (1989). Treatment of chronic hepatitis C with recombinant interferon alfa. *New England Journal of Medicine, 321,* 1501-1506.

13. Lemon, S. M., & Thomas, D. L. (1997). Drug therapy: Vaccines to prevent viral hepatitis *The New England Journal of Medicine, 336* , 196-204

14. Penn, F. (1993, October 21). Cancer confusion: The risks and realities of human papilloma virus. *Manhattan Spirit,* pp. 14-15; Rando, R. F. (1988). Human papillomavirus: Implications for clinical medicine. *Annals of Internal Medicine, 108,* 628-630.

15. Blakeslee, S. (1992, January 22). An epidemic of genital warts raises concern but not alarm. *The New York Times,* p. C12; Ochs, R. (1994, January 11). Cervical cancer comeback. *New York Newsday,* pp. 55, 57.

16. Koutsky, L. A., et al. (1992). A cohort study of the risk of cervical intraepithelial neoplasia Grade 2 or 3 in relation to papillomarvirus infection. *New England Journal of Medicine, 327,* 1272.

17. Penn (1993)

18. Blakeslee (1992)

19. Blakeslee (1992)

20. Cannistra, S. A., & Niloff, J. M. (1996). Cancer of the uterine cervix. *New England Journal of Medicine, 334,* 1030-1038.

21. Reinisch (1990)

22. Reinisch (1990)

23. Blakeslee (1992)

24. Blakeslee (1992)

25. Hatcher et al. (1990)

26. Blakeslee (1992)

27. Blakeslee (1992)

28. Penn (1993)

29. CDC (1989b)

30. CDC (1989b)

31. CDC (1989b)

32. Oriel, D. (1990). Genital human papillomavirus infection. In K. K. Holmes, et al. (Eds.), *Sexually transmitted diseases* (2nd ed.) (pp. 433-442). New York: McGraw-Hill, Inc.

### Chapter 12

1. Kunz & Finkel (1987)

2. Hatcher et al. (1990)

3. CDC (1989b); Hatcher et al. (1990)

4. Reinisch (1990)

5. CDC (1989b); Hatcher et al. (1990)

6. Reinisch (1990)

7. Reinisch (1990)

8. Levine (1991)

9. CDC (1989b)

10. Ronald, A. R., & Albritton, W. (1990). Chancroid and *Haemophilus ducreyi.* In K. K. Holmes, et al. (Eds.), *Sexually transmitted diseases* (2nd ed.) (pp. 263-272). New York: McGraw-Hill, Inc.

11. Ronald & Albritton (1990)

12. CDC (1989a). Summaries of identifiable diseases in the United States.

13. CDC (1989b)

14. Douglas, J. M., Jr. (1990). Molluscum contagiosum. In K. K. Holmes, et al. (Eds.), *Sexually*

*transmitted diseases* (2nd ed.) (pp. 443-448). New York: McGraw-Hill, Inc.

## Chapter 13

1.  Centers for Disease Control and Prevention. (1993). Sexually transmitted diseases treatment guidelines. *Morbidity and Mortality Weekly Report, 42*(No. RR-14)
2.  Mertz (1990); McCormack, W. M. (1990). Overview. *Sexually Transmitted Diseases, 57,* 187-191.
3.  Craig, M. E., Kalichman, S. C., & Follingstad, D. R. (1989). Verbal coercive sexual behavior among college students. *Archives of Sexual Behavior, 18,* 421-434.
4.  Lane, K. E., & Gwartney-Gibbs, P. A. (1985). Violence in the context of dating and sex. *Journal of Family Issues, 6,* 45-59.
5.  Lane & Gwartney-Gibbs (1985), p. 56
6.  Powell, E. (1991). *Talking back to sexual pressure.* Minneapolis: CompCare Publishers.
7.  Powell (1991)
8.  Items adapted from Powell, 1991, pp. 58-59, 70-75.
9.  Powell (1991), p. 72
10. Powell (1991), p. 72
11. Hunt, M. (1974). *Sexual behavior in the 1970's.* New York: Dell Books; Tavris, C., & Sadd, S. (1977). *The Redbook report on female sexuality.* New York: Delacorte.
12. Person, E. S., et al. (1989). Gender differences in sexual behaviors and fantasies in a college population. *Journal of Sex and Marital Therapy, 15,* 187-198.
13. Tavris & Sadd (1977)
14. CDC (1988b)
15. Reinisch (1990)
16. Conant, M. et al. (1986). Condoms prevent transmission of AIDS-associated retrovirus. *Journal of the American Medical Association, 255,* 1706. (Letter); Heiman, J. R., & LoPiccolo, J. (1987). *Becoming orgasmic* (2nd ed.). Englewood Cliffs, NJ: Prentice-Hall Books; Koop. C. E. (1988). *Understanding AIDS.* HHS Publication No. HHS—88—8404. Washington, D.C.: U.S. Government Printing Office.
17. CDC (1988b)
18. Carey, et al. (1992). Effectiveness of latex condoms as a barrier to human immunodeficiency virus-sized particles under conditions of simulated use. *Sexually Transmitted Diseases, 19,* 230-234.
19. Rosenberg, M. J., Hill, H. A., & Friel, P. J. (1991, October). *Spermicides and condoms in prevention of sexually transmitted diseases: A meta-analysis.* Paper presented at the meeting of the International Society for Sexually Transmitted Disease Research, Banff, Canada.
20. Rosenberg, M. J., & Gollub, E. L. (1992). Commentary: Methods women can use that may prevent sexually transmitted disease, including HIV. *American Journal of Public Health, 82,* 1473-1478.
21. CDC (1993a)
22. Gordon & Snyder (1989); Reinisch (1990)
23. Reinisch (1990)
24. Reinisch (1990)
25. Bell, R. (1980). *Changing bodies, changing lives.* New York: Random House.
26. Buhaug, H., et al. (1990). Should asymptomatic patients be tested for Chlamydia trachomatis in general practice? *British Journal of General Practice, 40,* 142-145.
27. Parra, W., et al. (1990). Patients counseling and behavior modification. In K. K. Holmes, et al. (Eds.), *Sexually transmitted diseases* (2nd ed.) (pp. 1057-1068). New York: McGraw-Hill, Inc.

## Chapter 14

1.  Hatcher et al. (1990); Reinisch (1990)
2.  Conant et al. (1986); Heiman & LoPiccolo (1987); Koop (1988)
3.  American Social Health Association (1991)
4.  Hein, K., DiGeronimo, T. F., & the Editors of Consumer Reports Books. (1989). *AIDS: Trading fears for facts.* Mount Vernon, NY: Consumers Union.
5.  Reinisch (1990)
6.  Reinisch (1990)
7.  American Social Health Association. (1989). *Condoms, contraceptives, and sexually transmitted diseases.* Research Triangle Park, NC: Author.
8.  Goldsmith, M. F. (1987). Sex in the age of AIDS calls for common sense and condom sense. *Journal of the American Medical Association, 257,* 2261-2266.
9.  Center for Disease Control and Prevention (1993). Sexually transmitted diseases treatment guidelines. MMWR 1993; 42 (No.RR-14)
10. Adapted from the CDC pamphlet, *Condoms and sexually transmitted diseases. . . especially AIDS*; American Social Health Association (1989); and other sources.
11. Adapted from the CDC pamphlet, *Condoms and sexually transmitted diseases. . . especially AIDS*; American Social Health Association (1989)